SpringerBriefs in Operations Research

More information about this series at http://www.springer.com/series/11467

Pouya Baniasadi • Vladimir Ejov • Jerzy A. Filar
Michael Haythorpe

Genetic Theory
for Cubic Graphs

 Springer

Pouya Baniasadi
School of CSEM
Flinders University
Bedford Park, SA, Australia

Jerzy A. Filar
School of CSEM
Flinders University
Bedford Park, SA, Australia

Vladimir Ejov
School of CSEM
Flinders University
Bedford Park, SA, Australia

Michael Haythorpe
School of CSEM
Flinders University
Bedford Park, SA, Australia

ISSN 2195-0482   ISSN 2195-0504   (electronic)
SpringerBriefs in Operations Research
ISBN 978-3-319-19679-4   ISBN 978-3-319-19680-0   (eBook)
DOI 10.1007/978-3-319-19680-0

Library of Congress Control Number: 2015942520

Springer Cham Heidelberg New York Dordrecht London

Printed on acid-free paper

Springer International Publishing AG Switzerland is part of Springer Science+Business Media (www. springer.com)

Acknowledgements

The authors gratefully acknowledge helpful discussions with V.S. Borkar, B.D. McKay, C.E. Praeger and P. Zograf. The research in this manuscript was made possible by grants from the Australian Research Council, specifically by the Discovery grants DP0666632, DP0984470 and DP120100532.

Contents

Author Bios

Pouya Baniasadi is a doctoral student in mathematics at Flinders University, Australia. He is a recipient of the AF Pillow Applied Mathematics Scholarship from the AF Pillow Mathematics Trust, Australia.

Vladimir Ejov is the leader of the Mathematical Analysis research group within Flinders Mathematical Sciences Laboratory, Flinders University, Australia. His research interests are in the areas of several complex variables, geometry of Cauchy-Riemann manifolds, combinatorial optimisation, Markov decision processes and graph theory.

Jerzy A. Filar is director of Flinders Mathematical Sciences Laboratory, Flinders University, Australia, and is a fellow of the Australian Mathematical Society. His research interests span both theoretical and applied topics in the fields of operations research, optimisation, game theory, applied probability and environmental modelling.

Michael Haythorpe is mathematician at Flinders University, Australia. His areas of research interest are numerical optimisation, computational mathematics, algorithm development and graph theory. He is a recipient of the AustMS Lift-off Fellowship Award 2010.

Chapter 1
Genetic Theory for Cubic Graphs

Abstract We partition the set of unlabelled cubic graphs into two disjoint sets, namely "genes" and "descendants", where the distinction lies in the absence or presence, respectively, of special edge cutsets. We introduce three special operations called breeding operations which accept, as input, two graphs, and output a new graph. The new graph inherits most of the structure of both the input graphs, and so we refer to the input graphs as parents and the output graph as a child. We also introduce three more operations called parthenogenic operations which accept a single descendant as input, and output a slightly more complicated descendant. We prove that every descendant can be constructed from a family of genes via the use of our six operations, and state the result (to be proved in Chap. 3) that this family is unique for any given descendant.

1.1 Introduction

A *graph* is an ordered pair $\Gamma = \langle V, E \rangle$ comprising a set of vertices V and a set of edges E, where entries of the latter take the form (i, j) for $i \in V$ and $j \in V$. Visually, graphs are often represented by drawing the entries of V as dots, with lines between vertices i and j whenever $(i, j) \in E$. For this reason, edges can be thought of as paths between vertices.

In general, graphs may contain *multiedges* (that is, entries (i, j) may appear in E with multiplicity greater than 1), or *loops* (that is, edges of the form (i, i)). Graphs that do not contain any multiedges or loops are called *simple*. Edges in a graph may also be *directed*, in the sense that a directed edge (i, j) describes a path existing from vertex i to vertex j, but not in reverse. If a graph has no directed edges, it is said to be *undirected*. Unless otherwise stated, every graph considered throughout this manuscript will be both simple and undirected. The number of edges incident on each vertex in an undirected graph is the *degree* of the vertex. If every vertex in the graph has the same degree, say k, the graph is said to be *k-regular*.

© Springer International Publishing Switzerland 2016
P. Baniasadi et al., *Genetic Theory for Cubic Graphs*, SpringerBriefs
in Operations Research, DOI 10.1007/978-3-319-19680-0_1

A *cycle*, or *closed walk*, is a sequence of vertices such that there is an edge incident on each adjacent pair of vertices in the cycle, and such that the first and last vertices in the sequence are the same. A *simple cycle* is one where no other repetition of vertices occurs in the sequence. In general, cycles are assumed to have length at least three, with cycles of length two called *trivial cycles*.

A particular graph may possess various properties that are important, and in many cases, determining the existence (or nonexistence) of such properties for a given graph is an interesting and difficult problem in its own right. Alternatively, graph-based problems may take the form of optimisation problems, where optimal structures within a graph are sought. Arguably, the most famous graph-based problem is the travelling salesman problem (TSP), in which each edge in a graph is assigned a weight, or cost. Then, the aim is to identify a subset of edges such that the total weight of selected edges is minimised, but that the edges chosen trace out a simple cycle that visits every vertex in the graph.

The line of research we describe in the present manuscript is motivated by the notion that, possibly, some of the underlying difficulty of challenging instances of the difficult graph-based problems is "inherited" from structures in simpler graphs which – in an appropriate sense – could be seen as "ancestors" of the given graph. For instance, could it be that the difficulty of a given instance of TSP is due to a structural complexity that arises already in a related smaller sized instance of an ancestor graph? If so, then this question is probably best studied in the context of the Hamiltonian cycle problem (HCP).

More precisely, the essence of HCP is contained in the following – deceptively simple – single sentence statement: *given a graph* $\Gamma = \langle V, E \rangle$, *find a simple cycle that contains all vertices* $v \in V$ *(Hamiltonian cycle (HC)) or prove that a HC does not exist*. With respect to this property – *Hamiltonicity* – graphs possessing HC are called Hamiltonian.[1]

The HCP is known to be NP-complete even when restricted to the class of *cubic*, that is, $3-regular$ graphs. Indeed, it is known that the class of cubic graphs is structurally very rich and has stimulated much research in its own right, including a number of still open conjectures (e.g., see Barnette [2] or Filar et al. [9]). Hence, this class has been selected for investigation in the present manuscript. In particular, we study the family $S := S(N)$ of undirected, unlabelled, connected cubic graphs on $N = 2m$ vertices, where m is a positive integer greater or equal to 2. We recall that a cubic graph has exactly three edges incident on every vertex. In our study we will be concerned with the problem of generating complex cubic graphs by appropriate compositions of simpler cubic graphs in a manner somewhat analogous to genetic "breeding". This will be achieved with the help of six "breeding operations".

Cubic graphs that cannot be seen as resulting from such operations will be called "genes", and all other cubic graphs will be called "descendants", with the latter vastly outnumbering the former. We will prove that any given descendant graph can

---

[1] The name stems from Sir William Hamilton's investigations of such cycles on the dodecahedron graph around 1856 but Leonhard Euler studied the famous "knight's tour" on a chessboard in this context as early as 1759.

be constructed from a finite family of genes. Since it will be seen, by construction, that most of the structure in a descendant is inherited from the genes, many properties of those genes are inherited as well. This motivates the introduction of "inverse operations" that, ultimately, decompose a given descendant into a family of "ancestor genes". Such a decomposition could subsequently be used with other graph theory algorithms to improve their solving time. However, we begin with an exploration of properties inherited by descendants in Chap. 2. In the third and final chapter of this manuscript, we will prove that each descendant has exactly one family of ancestor genes.

Undirected cubic graphs have been extensively studied in the literature (e.g., see [1, 8, 11, 12]). In particular, there are now programs that enumerate all instances (up to an automorphism) of cubic graphs on N vertices (e.g., see [6, 16]). A significant line of research concerns the generation of cubic graphs with the help of an exponential generating function (e.g., see [5, 16, 19]). To the extent that, in this manuscript, we generate descendant cubic graphs, the present contribution is conceptually, but not methodologically, related to the preceding. Indeed, the primary motivation of the present theory is not graph construction or enumeration, but rather to provide a framework for the decomposition of individual graphs. In this light, the result in Chap. 3 that each graph may only be decomposed into a unique set of smaller graphs via our methods, perhaps, is quite appealing.

The construction of ancestors and descendants of cubic graphs could be seen as being related to the work of McKay and Royle [15]. However, a feature of our approach is that all ancestors and descendants remain in the class of cubic graphs, and collectively constitute the entire set of connected cubic graphs. This enables us to study those properties of cubic graphs that may be "inherited" from simpler cubic graphs. Furthermore, the construction methods are quite different from those in [15]. Another similar investigation was by Tutte [16] who explored the construction of all 3-connected graphs (including 3-connected cubic graphs) from base graphs he called wheels. The work of Batagelj [3] is also similar to ours in that it is concerned with the generation of complex cubic graphs from simpler cubic graphs, where the former maintain the properties of the latter. In fact, some of the generating rules given in [3] are analogous to some of the breeding operations in this manuscript. However, the two approaches differ in that Batagelj's approach is to start with a single cubic graph and to then replace particular structures within the graph sequentially. In our approach, the majority of development occurs by combining multiple cubic graphs together, with each resulting graph being more complex than any of its ancestors.

1.2 Preliminaries

A standard concept in graph theory is that of *edge connectivity*. In the simple case, where the removal of a single, specific edge disconnects the graph, that edge is called a *bridge*. Extending this notion a set of edges *EC* constitutes an *edge cut* if their removal disconnects the graph. The smallest cardinality over all edge cuts in a given graph G is defined to be the *edge-connectivity* of that graph. Of course, cubic graphs can be k−edge-connected for only three values of k: $1, 2$ or 3.

We note, however, that in some 3−edge-connected cubic graphs (e.g., the famous Petersen graph, see Fig. 1.2) the removal of any minimal edge cut set isolates a single vertex. Arguably, such a partition of vertices is in a sense degenerate, and prevents a more refined classification of cubic graphs. To address this problem and achieve a finer classification we introduce a special class of edge cut sets that we name *crackers*. The latter are defined as follows.

Definition 1.1. An edge cut set Γ consisting of k edges in a cubic graph Γ is a *k-cracker* if no two edges of the cut set are adjacent in the sense of being incident on the same vertex, and no proper subset of these edges disconnects the graph.

It is important to note that, according to Definition 1.1, crackers are only defined in the context of cubic graphs. Although obvious generalisations exist for other types of graphs, forthcoming results in this manuscript about crackers will assume cubicity.

Lemma 1.1. *The removal of a cracker from a connected cubic graph results in exactly two connected components.*

Proof. Since any cracker is an edge cut set, the removal of a cracker must result in at least two connected components. Assume there exists a cracker $C = \{e_1, \ldots, e_k\}$ whose removal results in more than two connected components. We can think of the removal of C as a sequence of individual edge removals, for each edge in C. It is clear that the removal of a single edge cannot result in more than one additional disjoint component. Therefore, after removing edges $e_1, \ldots, e_{k-1}$, but before removing edge e_k, there must already be at least two connected components. However, this implies that a proper subset of C also disconnects the graph, which violates Definition 1.1. Therefore, C is not a cracker, and the initial assumption is false. □

Of course, any cracker is not only an edge cut set, but in fact a *cyclic edge cut set*. This is clear because its removal disconnects the graph into two connected subgraphs, and cubicity ensures that both of these subgraphs must contain at least three vertices. Since each remaining vertex has degree at least two (as only non-adjacent edges were removed), these connected subgraphs must contain cycles. It is then clear that a minimal cyclic edge cut set in a given cubic graph is a cracker of minimal cardinality for that graph. However, larger cyclic edge cut sets may contain adjacent edges, and therefore not all cyclic edge cut sets are crackers.

We recall that the *girth*, g, of a graph is the length of the shortest (nontrivial) cycle in the graph. In many cases, a nontrivial cycle determining a cubic graph's girth automatically induces a g-cracker made up of edges that are not in the cycle but which are incident on one vertex of the cycle. However, many graphs have crackers of size less than g. For instance, it is easy to see that for the graph given in Fig. 1.1, $g = 3$ but the edges e_1 and e_2 form a 2-cracker.

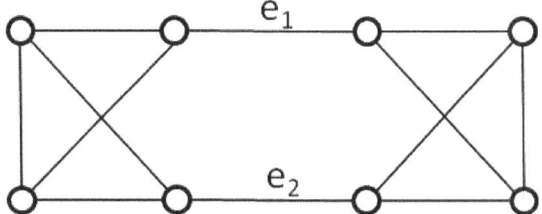

Fig. 1.1 An 8-vertex cubic graph with girth 3 and the 2-cracker $\{e_1, e_2\}$

We note also that there are only two connected cubic graphs, on 4 and 6 vertices respectively, that contain no crackers at all (see Fig. 1.4 and the discussion in Sect. 1.3).

Remark 1.1. Note that, since any minimal cyclic edge cut set in a cubic graph is a cracker, it is clear that for any given cubic graph, the cyclic edge connectivity is equal to the size of the smallest cracker in that graph. For the sake of simplifying the notation, we will refer to a cyclically k-connected graph as a C_k-connected graph. The class of all C_k-connected cubic graphs on N vertices will henceforth be denoted by $S_k(N)$, or simply by S_k, when the number of vertices is fixed.

We note that the famous Petersen graph is C_5-connected in the above sense (see Fig. 1.2), as the edges connecting the "inner-star" to the outer boundary form one of a number of 5−crackers, and no smaller crackers exist in this graph.

1.3 Motivation

It follows immediately from Remark 1.1 that (for fixed $N \geq 8$) the class $S(N)$ of connected cubic graphs on N vertices can be partitioned as

$$S(N) = \bigcup_{i=1}^{M} S_i,$$

where $M \leq \dfrac{N}{2}$. The choice of upper bound is conservative because at most $\dfrac{N}{2}$ non-adjacent edges can be chosen in any graph of size N, but in reality it is likely that far fewer than $\dfrac{N}{2}$ partitions will be required for any given N. For instance, it can be confirmed by exhaustive search that when $N = 20$, $M = 6$. However, it should be noted that the bound on M is tight; there are cubic graphs with N vertices whose smallest cracker is of size $\frac{N}{2}$, such as the 8-vertex graph displayed in Fig. 1.3.

Definition 1.2. If a cubic graph Γ is C_k-connected for $k \geq 4$, we call it a *gene*. Otherwise, we call Γ a *descendant*.

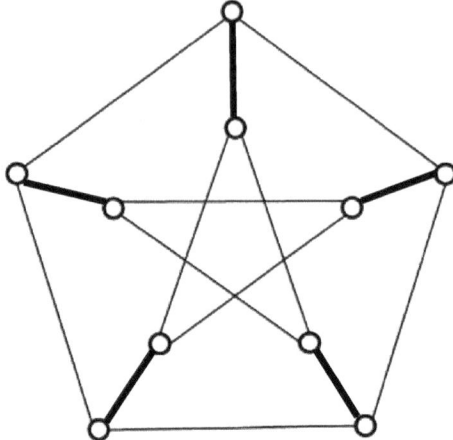

Fig. 1.2 Petersen graph, which is C_5-connected. The thick edges highlight one of the 5-crackers in the graph

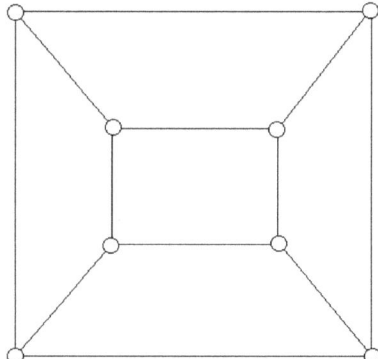

Fig. 1.3 A C_4-connected graph containing 8 vertices

The reasoning behind the choice of names *gene* and *descendant* is made clear in Sect. 1.4, where we demonstrate that any descendant can be obtained from a finite family of genes, through the use of prescribed *breeding operations* that introduce crackers into a descendant. Numerical testing has shown that genes are far less numerous than descendants.

It was mentioned earlier that two cubic graphs, namely the 4-vertex gene Γ_4^* and the 6-vertex gene Γ_6^*, contain no crackers at all. These two graphs can be seen in Fig. 1.4. The following lemma proves that every other cubic graph contains at least one cracker, and that the size of the smallest cracker is bounded above by the girth of the graph.

Lemma 1.2. *Except for Γ_4^* and Γ_6^*, all connected cubic graphs contain at least one cracker of size no more than the girth g of the graph.*

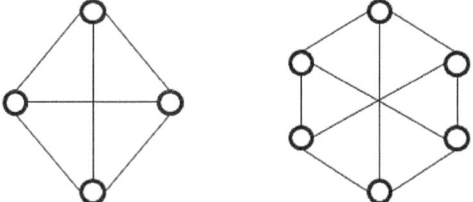

Fig. 1.4 Cubic graphs Γ_4^* and Γ_6^*, which contain no crackers

Proof. The cases of Γ_4^* and Γ_6^* can be confirmed by inspection. There is one other cubic graph containing 6 vertices, with girth 3, which contains a 3-cracker, as displayed in Fig. 1.5, and Γ_4^* is the only cubic graph containing 4 vertices. So the lemma is true for $N < 8$.

For any cubic graph containing 8 or more vertices, it was proved in Lou et al. [14] that there exists at least one cyclic edge cut set of size g. The smallest cyclic edge cut set in a cubic graph is a cracker, so it is clear that the smallest cracker can be of size no bigger than g. $\quad\square$

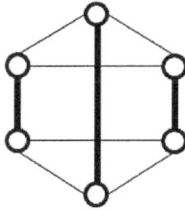

Fig. 1.5 A 6-vertex cubic graph, with girth 3, which contains a 3-cracker comprising the highlighted edges

In Sect. 1.5, it will be proved that any descendant graph can be obtained from a finite family of genes. Furthermore, in Chap. 3, it will be proved that this family of genes is unique for any given descendant. These results motivate the second chapter of this manuscript, in which we observe that descendants inherit many of their properties from the genes used to construct them. In these cases, analysis of a descendant can be reduced to the problem of analysing the component genes, which are often much smaller than the descendant.

One such graph theoretic property, investigated in detail in Chap. 2, is the aforementioned property of *Hamiltonicity*, that is, the property of containing a simple cycle of length equal to the number of vertices in the graph. We observe that non-Hamiltonian genes are extremely rare. Even excluding the (trivially non-Hamiltonian) *bridge graphs*, which are descendants by definition, non-Hamiltonian descendants constitute a large majority of the remaining non-Hamiltonian graphs; see Table 1.1. The second column of that table, labelled by NH_1, lists the percentages of bridge graphs relative to the total cardinality of non-Hamiltonian graphs,[2] denoted by NH. The third column labelled by $NH_{2+} := NH \setminus NH_1$, lists the percentages of graphs which are C_k-connected for $k \geq 2$, relative to the cardinality of NH. The fourth column labelled by NH_{4+}, lists the percentages of graphs which are C_k-connected for $k \geq 4$, relative to the cardinality of NH. Finally, the fifth column labelled by NH_{4+}/NH_{2+}, lists the percentages of graphs which are C_k-connected for $k \geq 4$, relative to the cardinality of all non-bridge, non-Hamiltonian, graphs in NH_{2+}.

We shall define graphs in NH_{4+} as *mutants*, a name that properly reflects their exceptionality. For instance, we note from the fourth column of Table 1.1 that with $N = 18$ only 0.12 of one percent of non-Hamiltonian graphs are mutants, which corresponds to two (out of 1666) non-Hamiltonian cubic graphs on 18 vertices. These two mutants are the famous Blanuša Snarks [4]. Of course, mutants are simply those non-Hamiltonian, cubic, graphs that do not possess any cubic crackers.

In fact, it is no coincidence that the Blanuša Snarks appear as mutants in this framework. In Read and Wilson [18], the definition of an irreducible Snark is given as a cubic graph with edge chromatic number of 4, girth 5 or more, and not containing three edges whose deletion results in a disconnected graph, each of whose components is nontrivial. This final condition, along with the well known fact that all Snarks are non-Hamiltonian, is akin to our definition of a mutant. Therefore, the set of all mutants is a superset of the set of all irreducible Snarks. However, some non-Snark mutants do exist, and therefore have an edge chromatic number of 3, and possibly girth 4 as well. In particular, the BH-Mutant on 20 vertices, displayed in Fig. 1.6 is the smallest non-Snark mutant. There are 16 further non-Snark mutants of size 22, one of which is the Zircon-Mutant, also displayed in Fig. 1.6.

1.4 Breeding and Parthenogenic Operations

From Definition 1.2, it is clear that genes do not contain any 1-crackers, 2-crackers, or 3-crackers. Collectively, we refer to 1-crackers, 2-crackers and 3-crackers as *cubic crackers*. As a corollary, descendants must contain at least one cubic cracker. It then seems plausible that we might be able to construct any given descendant by combining two or more cubic graphs together in such a fashion as to introduce the cubic crackers present in that descendant graph. Since there are three different types of cubic crackers, we define three *breeding operations* that map two cubic graphs to a single descendant by inserting a cubic cracker between them in such a fashion as to retain cubicity. In such a case, we say that the descendant has been obtained by *breeding*. We refer to the original two cubic graphs as the *parents* of the descendant graph, and likewise the descendant graph is the *child* of the two parents.

---

[2] For a more complete study of the prevalence of cubic bridge graphs relative to the total set of cubic non-Hamiltonian graphs, see Filar et al. [9].

	NH$_1$ (%)	NH$_{2+}$ (%)	NH$_{4+}$ (%)	NH$_{4+}$/NH$_{2+}$ (%)
10 Vertex	50.00	50.00	50.00	100
12 Vertex	80.00	20.00	0	0
14 Vertex	82.86	17.14	0	0
16 Vertex	84.93	15.07	0	0
18 Vertex	86.13	13.86	0.12	0.86
20 Vertex	87.40	12.60	0.05	0.38
22 Vertex	88.59	11.41	0.02	0.21

Table 1.1 Distribution of Non-Hamiltonian (NH) C_k-connected graphs

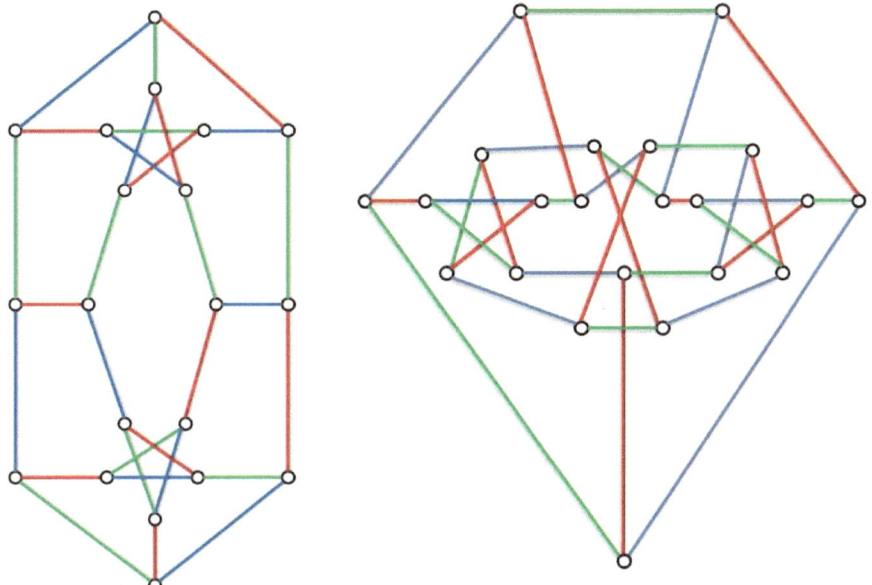

Fig. 1.6 BH-Mutant which is a girth 5, C_4-connected non-Hamiltonian non-Snark mutant containing 20 vertices, and Zircon-Mutant which is a girth 5, C_5-connected non-Hamiltonian non-Snark mutant containing 22 vertices

Note that the following operations are defined only for cubic graphs. In addition, although the definitions are still valid for disconnected cubic graphs, in this chapter we are interested only in connected cubic graphs, and make the assumption that all input graphs are connected and cubic.

1.4.1 Breeding Operations

Definition 1.3. A *type 1 breeding operation* is a function $\mathscr{B}_1$ defined on the tuple $(\Gamma_1, \Gamma_2, e_1, e_2)$, where $\Gamma_1 = \langle V_1, E_1 \rangle$ and $\Gamma_2 = \langle V_2, E_2 \rangle$ are cubic graphs, and furthermore, $e_1 = (a,b) \in E_1$ and $e_2 = (c,d) \in E_2$. This function maps such a tuple onto another tuple (Γ_D, e) as follows

$$\mathscr{B}_1(\Gamma_1, \Gamma_2, e_1, e_2) = (\Gamma_D, e),$$

where $\Gamma_D = \langle V_D, E_D \rangle$ and $\{e = (v_1, v_2)\} \in E_D$. The new set of vertices is $V_D = V_1 \cup V_2 \cup v_1 \cup v_2$. The new set of edges is $E_D = (E_1 \backslash e_1) \cup (E_2 \backslash e_2) \cup \{(a,v_1), (b,v_1), (c,v_2), (d,v_2), (v_1,v_2)\}$.

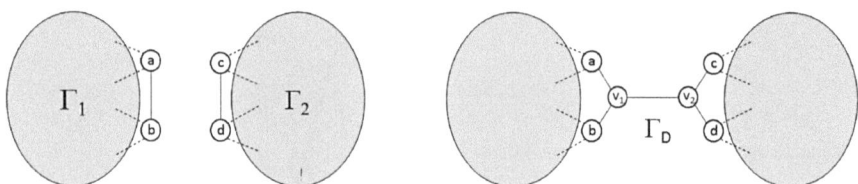

Fig. 1.7 Graphs Γ_1, Γ_2 and Γ_D as described in Definition 1.3

Note that a type 1 breeding operation always outputs a *bridge graph* (that is, a C1-connected graph). See Fig. 1.7 for an illustration.

Definition 1.4. A *type 2 breeding operation* is a function $\mathscr{B}_2$ defined on the tuple $(\Gamma_1, \Gamma_2, a, b, c, d)$, where $\Gamma_1 = \langle V_1, E_1 \rangle$ and $\Gamma_2 = \langle V_2, E_2 \rangle$ are cubic graphs, and furthermore, $e_1 = (a,b) \in E_1$ and $e_2 = (c,d) \in E_2$ and neither edge is a 1-cracker. This function maps such a tuple onto another tuple (Γ_D, e_3, e_4) as follows

$$\mathscr{B}_2(\Gamma_1, \Gamma_2, a, b, c, d) = (\Gamma_D, e_3, e_4),$$

where $\Gamma_D = \langle V_D, E_D \rangle$ and $\{e_3 = (a,c), e_4 = (b,d)\} \in E_D$. The new set of vertices is $V_D = V_1 \cup V_2$. The new set of edges is $E_D = (E_1 \backslash e_1) \cup (E_2 \backslash e_2) \cup \{(a,c), (b,d)\}$.

Clearly Γ_D contains the 2-cracker $\{(a,c), (b,d)\}$. See Fig. 1.8 for an illustration. Note also that a type 2 breeding operation always creates a 2-edge-connected descendant, unless either of Γ_1 or Γ_2 is 1-edge-connected (in which case, Γ_D is also 1-edge-connected).

Definition 1.5. A *type 3 breeding operation* is a function $\mathscr{B}_3$ defined on the tuple $(\Gamma_1, \Gamma_2, v_1, v_2, a, b, c, d, e, f)$, where $\Gamma_1 = \langle V_1, E_1 \rangle$ and $\Gamma_2 = \langle V_2, E_2 \rangle$ are cubic graphs, and furthermore, $v_1 \in V_1$ is incident to vertices a, b and $c \in V_1$ and $v_2 \in V_2$ is incident to vertices d, e and $f \in V_2$. None of the edges adjacent to v_1 or v_2 are 1-crackers. This function maps such a tuple onto another tuple $(\Gamma_D, e_1, e_2, e_3)$ as follows

$$\mathscr{B}_3(\Gamma_1, \Gamma_2, v_1, v_2, a, b, c, d, e, f) = (\Gamma_D, e_1, e_2, e_3),$$

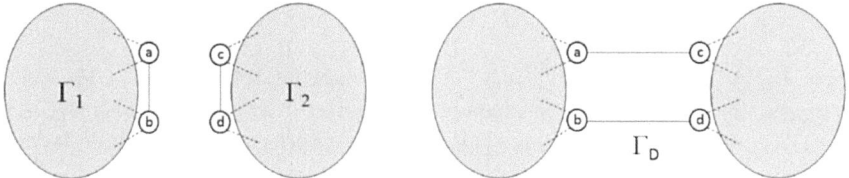

Fig. 1.8 Graphs Γ_1, Γ_2 and Γ_D as described in Definition 1.4

where $\Gamma_D = \langle V_D, E_D \rangle$, and also, $\{e_1 = (a,d), e_2 = (b,e), e_3 = (c,f)\} \in E_D$. The new set of vertices is $V_D = (V_1 \backslash v_1) \cup (V_2 \backslash v_2)$. The new set of edges is $E_D = (E_1 \backslash \{(v_1,a),(v_1,b),(v_1,c)\}) \cup (E_2 \backslash \{(v_2,d),(v_2,e),(v_2,f)\}) \cup \{(a,d),(b,e),(c,f)\}$.

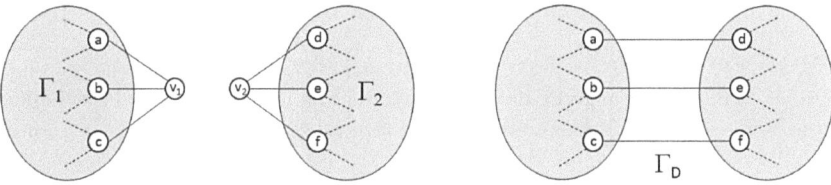

Fig. 1.9 Graphs Γ_1, Γ_2 and Γ_D as described in Definition 1.5

See Fig. 1.9 for an illustration of type 3 breeding. Note that the order of vertices in the definitions of types 2 and 3 breeding is important, as it determines which edges appear in the descendant. So, in general, $\mathscr{B}_2(\Gamma_1, \Gamma_2, a, b, c, d) \neq \mathscr{B}_2(\Gamma_1, \Gamma_2, b, a, c, d)$, and likewise for type 3 breeding.

1.4.2 Parthenogenic Operations

In addition to the preceding three breeding operations, we also define three *parthenogenic operations*. These are operations that map a single descendant to a new, more complex, descendant by replacing a cracker in the original descendant with two new crackers. We say that such a new descendant has been obtained from *parthenogenesis*. For simplicity of terminology, we will again refer to the original descendant as the *parent* of the new descendant, and likewise we will refer to the new descendant as the *child* of the original descendant. Also for simplicity of terminology, we will refer to the three breeding operations and the three parthenogenic operations collectively as the six *breeding operations*.

Definition 1.6. A *type 1 parthenogenic operation* is a function $\mathscr{P}_1$ defined on the tuple (Γ_1, e_1) where $\Gamma_1 = \langle V_1, E_1 \rangle$ is a bridge graph and $e_1 = (a,b) \in E_1$ is a 1-cracker. This function maps such a tuple onto another tuple (Γ_D, v_1, v_4) as follows

$$\mathscr{P}_1(\Gamma_1, e_1) = (\Gamma_D, v_1, v_4),$$

where $\Gamma_D = \langle V_D, E_D \rangle$ and $\{v_1, v_4\} \in V_D$. The new set of vertices is $V_D = V_1 \cup \{v_1, v_2, v_3, v_4\}$. The new set of edges is $E_D = (E_1 \backslash e_1) \cup \{(a, v_1), (v_1, v_2), (v_1, v_3), (v_2, v_3), (v_2, v_4), (v_3, v_4), (v_4, b)\}$. This process inserts an additional 1-cracker into Γ_D.

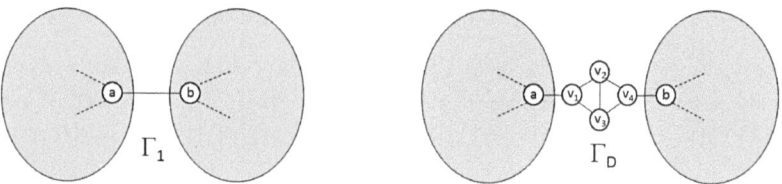

Fig. 1.10 Graphs Γ_1 and Γ_D as described in Definition 1.6

We refer to the subgraph $\Gamma_S = \langle \{v_1, v_2, v_3, v_4\}, \{(v_1, v_2), (v_1, v_3), (v_2, v_3), (v_2, v_4), (v_3, v_4)\} \rangle$ as the *parthenogenic diamond*, Fig. 1.11 and say that a type 1 parthenogenic operation inserts a parthenogenic diamond into a bridge. See Fig. 1.10 for an illustration.

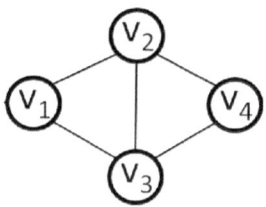

Fig. 1.11 A parthenogenic diamond

Definition 1.7. A *type 2 parthenogenic operation* is a function $\mathscr{P}_2$ defined on the tuple (Γ_1, e_1, e_2) where $\Gamma_1 = \langle V_1, E_1 \rangle$ is a cubic graph containing a 2-cracker comprising two edges $e_1 = (a, b)$ and $e_2 = (c, d)$. This function maps such a tuple onto another tuple (Γ_D, v_1, v_2) as follows

$$\mathscr{P}_2(\Gamma_1, e_1, e_2) = (\Gamma_D, v_1, v_2),$$

where $\Gamma_D = \langle V_D, E_D \rangle$ and $\{v_1, v_2\} \in V_D$. The new set of vertices is $V_D = V_1 \cup \{v_1, v_2\}$. The new set of edges is $E_D = (E_1 \backslash \{e_1, e_2\}) \cup \{(a, v_1), (b, v_1), (c, v_2), (d, v_2), (v_1, v_2)\}$. This process inserts an additional 2-cracker into Γ_D.

We refer to the subgraph $\Gamma_S = \langle \{v_1, v_2\}, (v_1, v_2) \rangle$ as the *parthenogenic bridge*, Fig. 1.13 and say that a type 2 parthenogenic operation inserts a parthenogenic bridge into a 2-cracker. See Fig. 1.12 for an illustration.

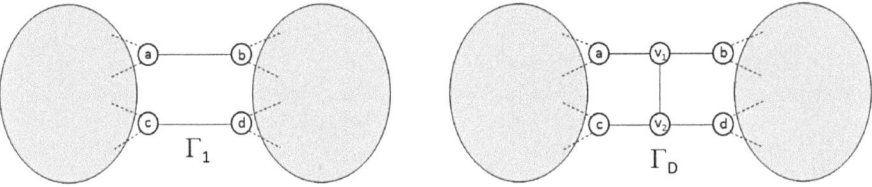

Fig. 1.12 Graphs Γ_1 and Γ_D as described in Definition 1.7

Fig. 1.13 A parthenogenic bridge

Definition 1.8. A *type 3 parthenogenic operation* is a function $\mathscr{P}_3$ defined on the tuple (Γ_1, a) where $\Gamma_1 = \langle V_1, E_1 \rangle$ is a bridge graph and $a \in V_1$ is a vertex incident to a 1-cracker composing an edge $e_1 = (a, b) \in E_1$ and is adjacent to vertices c and $d \in V_1$. This function maps such a tuple onto another tuple (Γ_D, a, v_1, v_2) as follows

$$\mathscr{P}_3(\Gamma_1, a) = (\Gamma_D, a, v_1, v_2),$$

where $\Gamma_D = \langle V_D, E_D \rangle$ and $\{a, v_1, v_2\} \in V_D$. The new set of vertices is $V_D = V_1 \cup \{v_1, v_2\}$. The new set of edges is $E_D = (E_1 \backslash \{(a, c), (a, d)\}) \cup \{(a, v_1), (a, v_2), (v_1, v_2), (v_1, c), (v_2, d)\}$.

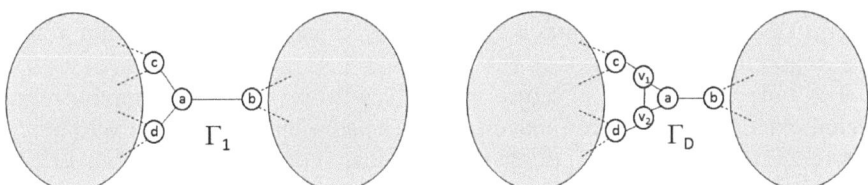

Fig. 1.14 Graphs Γ_1 and Γ_D as described in Definition 1.8

We refer to the subgraph $\Gamma_S = \langle \{a, v_1, v_2\}, \{(a, v_1), (a, v_2), (v_1, v_2)\} \rangle$ as the *parthenogenic triangle*, Fig. 1.15 and say that a type 3 parthenogenic operation inserts such a parthenogenic triangle next to the 1-cracker. See Fig. 1.14 for an illustration.

The following result is nearly self-evident form construction, however, a proof is included below for the sake of completeness.

Lemma 1.3. *A child graph resulting from any of the six breeding operations is connected.*

Proof. The nature of the six breeding operations is that the parent graphs are mostly unaltered, and are changed only in a neighbourhood of the introduced cubic cracker.

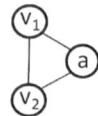

Fig. 1.15 A parthenogenic triangle

Therefore, we can focus just on these areas. Since the parent graphs are (by assumption) connected to begin with, we only need to be concerned with which edges, present in the parent graphs, are not present in the child graph.

For the cases of type 2 breeding, and types 1 and 2 parthenogenesis, only a single edge from the parent graph (or from each of the parent graphs in the case of type 2 breeding) is missing in the child graph. In all situations other than type 1 parthenogenesis, we see by definition that the missing edge cannot be a 1-cracker, and therefore, the parent graphs remain connected. By construction it is clear that the adjoining cracker ensures the child graph is also connected. In the case of type 1 parthenogenesis, once the 1-cracker is removed separating the graph into two connected components, the construction immediately creates a new connected subgraph that is joined to both components. Since no other edges or vertices are removed, the child graph is therefore connected.

For type 1 breeding, only a single edge is removed from each parent graph. If neither edge is a 1-cracker, then the argument in the previous paragraph can be used to show the child graph is connected. However, it is possible that one or both removed edges could be 1-crackers. If so, the corresponding parent graphs become disconnected. However, if this is the case, the graphs are reconnected by the introduction of vertices v_1 and v_2 (see Definition 1.3). Indeed, triplets of vertices (a, v_1, b) and (c, v_2, d) perform the connecting function of the removed edges (a, b) and (c, d). It is then clear by construction that the adjoining cracker ensures the child graph is also connected.

For type 3 breeding, a vertex is removed from both parent graphs. From Definition 1.5, we know that none of the edges adjacent to these two vertices constitute 1-crackers. Therefore, the removal of these vertices cannot disconnect either graph. By construction it is then clear that the adjoining cracker ensures the child graph is also connected.

Finally, for type 3 parthenogenesis, although two (adjacent) edges from the parent graph are missing in the child graph, it is clear from the latter's construction that this can not result in a disconnected descendant. □

1.4.3 Inverse Breeding and Inverse Parthenogenic Operations

For some tuples (Γ_D, e) where $\{e\} \in E_D$ is a 1-cracker, the inverse breeding operation $\mathscr{B}_1^{-1}(\Gamma_D, e) = (\Gamma_1, \Gamma_2, e_1, e_2)$ is automatically well defined. In such a case $\{e\}$ is called an *irreducible* 1-cracker. If not, $\{e\}$ will be called a *reducible* 1-cracker.

Similarly if the inverse operation $\mathcal{B}_2^{-1}(\Gamma_D, e_3, e_4) = (\Gamma_1, \Gamma_2, e_1, e_2)$ is well defined, where $\{e_3, e_4\} \in E_D$ is a 2-cracker, the 2-cracker is called an irreducible 2-cracker and reducible 2-cracker otherwise. We will show later that the inverse operation $\mathcal{B}_3^{-1}(\Gamma_D, e_1, e_2, e_3) = (\Gamma_1, \Gamma_2, v_1, v_2, a, b, c, d, e, f)$ is always defined where $\{e_1, e_2, e_3\} \in E_D$ is a 3-cracker. Therefore every 3-cracker is irreducible.

Definition 1.9. Whenever a cubic cracker is irreducible one of Eqs. (1.1)–(1.3) defines the corresponding inverse breeding operation $\mathcal{B}_1^{-1}(.)$, $\mathcal{B}_2^{-1}(.)$ or $\mathcal{B}_3^{-1}(.)$. The two cubic graphs (Γ_1, Γ_2) from the tuple produced by these operations are parents of Γ_D. In particular,

$$\mathcal{B}_1^{-1}(\Gamma_D, e) = (\Gamma_1, \Gamma_2, e_1, e_2) \tag{1.1}$$

where Γ_D, e, Γ_1, Γ_2, e_1 and e_2 are defined in Definition 1.3. Similarly,

$$\mathcal{B}_2^{-1}(\Gamma_D, e_3, e_4) = (\Gamma_1, \Gamma_2, e_1, e_2) \tag{1.2}$$

where Γ_D, e_3, e_4, Γ_1, Γ_2, e_1 and e_2 are defined in Definition 1.4. Also,

$$\mathcal{B}_3^{-1}(\Gamma_D, e_1, e_2, e_3) = (\Gamma_1, \Gamma_2, v_1, v_2, a, b, c, d, e, f) \tag{1.3}$$

where Γ_D, e_1, e_2, e_3, Γ_1, Γ_2, v_1, v_2, a, b, c, d, e and f are defined in Definition 1.5.

Similarly, inverse parthenogenic operations can be defined as follows.

Definition 1.10. Equations (1.4)–(1.6) define the corresponding inverse partheno genic operations $\mathcal{P}_1^{-1}(.)$, $\mathcal{P}_2^{-1}(.)$ or $\mathcal{P}_3^{-1}(.)$. The cubic graph (Γ_1) from the tuple produced by these operations is a parent of Γ_D. In particular,

$$\mathcal{P}_1^{-1}(\Gamma_D, v_1, v_4) = (\Gamma_1, e_1), \tag{1.4}$$

where Γ_D, v_1, v_4, Γ_1 and e_1 are defined in Definition 1.6. Similarly,

$$\mathcal{P}_2^{-1}(\Gamma_D, v_1, v_2) = (\Gamma_1, e_1, e_2), \tag{1.5}$$

where Γ_D, v_1, v_2, Γ_1, e_1 and e_2 are defined in Definition 1.7. Also,

$$\mathcal{P}_3^{-1}(\Gamma_D, a, v_1, v_2) = (\Gamma_1, a), \tag{1.6}$$

where Γ_D, a, Γ_1, Γ_2, v_1 and v_2 are defined in Definition 1.8.

Collectively, we refer to the three inverse breeding operations and the three inverse parthenogenic operations as the six *inverse operations*.

It is important to note that the six breeding operations and the six inverse operations presented here are not entirely new, and have been used in various forms in other cubic graph generation routines. For example, type 2 parthenogenesis induces an *H-subgraph* which is well-studied in literature (e.g. see Ore [17]). Type 1 inverse parthenogenesis and type 1 inverse breeding appear as Operation $O_2(K_4)$ and Operations $\mathcal{R}$ respectively in Ding and Kanno [7]. Types 1 and 2 parthenogenesis appear in Brinkmann [5]. All of the six breeding operations except type 3 breeding appear in some sense as generating rules in Batagelj [3], specifically generating rules P1, P2, P3, P4 an P8. However, all of the above works involve growing the complexity of a single graph by the evolution of subgraphs, rather than combining several cubic graphs together. In addition, the operations in the above works that are analogous to our parthenogenic operations are not confined to the same conditions as ours (that is, they must occur on 1-crackers and 2-crackers). Little consideration is given in the above works to procedures that are analogous to our inverse operations. The benefits and potency of the particular set of breeding and inverse operations that we have detailed in this section are demonstrated in the following two sections.

1.5 Results

The following three propositions relate to the different possible methods of introducing cubic crackers via the six breeding operations, and are used in the proof of the main theorem for this section, Theorem 1.1.

Proposition 1.1. *Any descendant involving a 1-cracker can be obtained by performing either type 1 breeding, type 1 parthenogenesis, or type 3 parthenogenesis on appropriate subgraph(s).*

Proof. Consider a descendant $\Gamma_D = \langle V_D, E_D \rangle$ containing a 1-cracker comprising an edge (v_1, v_2). Since Γ_D is cubic, v_1 and v_2 will both be adjacent to two more vertices, say $\{a, b\}$ and $\{c, d\}$ respectively. Since the 1-cracker comprises edge (v_1, v_2), we know that $\{a, b\}$ and $\{c, d\}$ are disjoint sets. Then, we can consider two cases.

Case 1: The edges (a, b) and (c, d) are not present in Γ_D. In this case, the 1-cracker is irreducible. Suppose the bridge (v_1, v_2) is removed from Γ_D, separating the graph into two sub-cubic subgraphs, $\Gamma_{S1} = \langle V_{S1}, E_{S1} \rangle$ and $\Gamma_{S2} = \langle V_{S2}, E_{S2} \rangle$, in which v_1 and $v-2$ each has degree 2. Without loss of generality, we assume that $v_1 \in V_{S1}$, and $v_2 \in V_{S2}$. Then, we define a cubic graph $\Gamma_1 = \langle V_1, E_1 \rangle$, where $V_1 = V_{S1} \backslash v_1$ and $E_1 = (E_{S1} \backslash \{(a, v_1), (b, v_1)\}) \cup (a, b)$. Similarly, we define a second cubic graph $\Gamma_2 = \langle V_2, E_2 \rangle$, where $V_2 = V_{S2} \backslash v_2$ and $E_2 = (E_{S2} \backslash \{(c, v_2), (d, v_2)\}) \cup (c, d)$. Then, Γ_D can be obtained from the type 1 breeding operation $\mathscr{B}_1(\Gamma_1, \Gamma_2, (a, b), (c, d))$. This situation can be seen in Fig. 1.16.

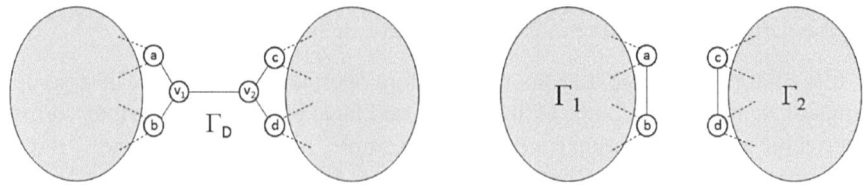

Fig. 1.16 Graphs Γ_D, Γ_1 and Γ_2 as described in Case 1 of Proposition 1.1

Case 2: At least one of the edges (a,b) or (c,d) is present in Γ_D. In this case, the 1-cracker cannot be obtained from a type 1 breeding operation, as such an operation would remove these edges. A 1-cracker of this type is, therefore, reducible. If both edges are present, we can focus on either one. Without loss of generality, we will assume that edge $(a,b) \in E_D$. Then, since Γ_D is cubic, and vertices a and b are both adjacent to vertex v_1 and to each other, they will also be adjacent to one more vertex each, say vertices e and f respectively. Note that it is possible that $e = f$, so we need to consider the cases separately.

Case 2.1: $e = f$. Then, both edges (a,e) and (b,e) are in E_D, as seen in the right panel of Fig. 1.17. Since Γ_D is cubic, vertex e is adjacent to a third vertex, say g. Clearly, edge (e,g) is a bridge, since edge (v_1,v_2) is a bridge. Therefore, Γ_D contains a parthenogenic diamond through vertices e,a,b,v_1. Then, we define a cubic graph $\Gamma_1 = \langle V_1, E_1 \rangle$, where $E_1 = (E_D \setminus \{(v_1,v_2),(a,v_1),(b,v_1),(a,b),(a,e),(b,e),(e,g)\}) \cup (g,v_2)$ and $V_1 = V_D \setminus \{v_1,a,b,e\}$. We can then obtain Γ_D from the type 1 parthenogenic operation $\mathscr{P}_1(\Gamma_1,(g,v_2))$. This situation can be seen in Fig. 1.17.

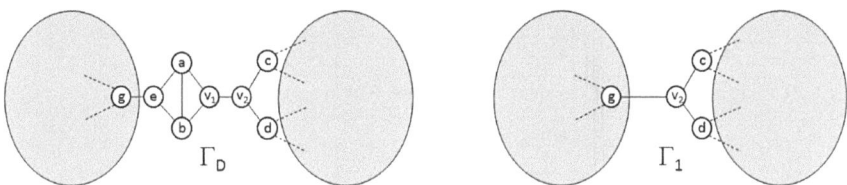

Fig. 1.17 Graphs Γ_D and Γ_1 as described in Case 2.1 of Proposition 1.1

Case 2.2: $e \neq f$. In this situation, Γ_D contains a parthenogenic triangle through vertices a,b,v_1. Then, as illustrated in Fig. 1.18, we define a cubic graph $\Gamma_1 = \langle V_1, E_1 \rangle$, where $E_1 = (E_D \setminus \{(a,v_1),(b,v_1),(a,b),(a,e),(b,f)\}) \cup \{(e,v_1),(f,v_1)\}$ and $V_1 = V_D \setminus \{a,b\}$. Then, we can obtain Γ_D from the type 3 parthenogenic operation $\mathscr{P}_3(\Gamma_1,v_1)$. This situation can be seen in Fig. 1.18. $\square$

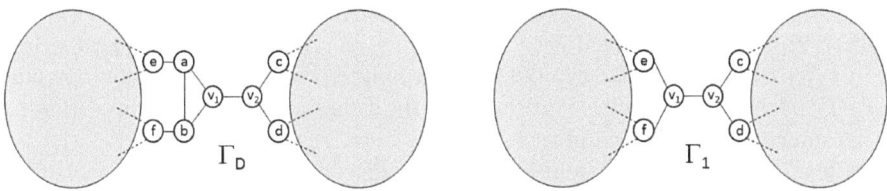

Fig. 1.18 Graphs Γ_D and Γ_1 as described in Case 2.2 of Proposition 1.1

Proposition 1.2. *Any descendant involving a 2-cracker can be obtained by performing either type 1 breeding, type 2 breeding, type 1 parthenogenesis, type 2 parthenogenesis, or type 3 parthenogenesis on appropriate subgraph(s).*

Proof. Consider a descendant $\Gamma_D = \langle V_D, E_D \rangle$ containing a 2-cracker comprising edges (v_1, v_2) and (v_3, v_4). We consider two cases.

Case 1: Neither edge (v_1, v_3) nor (v_2, v_4) are present in Γ_D, as illustrated in Fig. 1.19. In this case, the 2-cracker is irreducible. Suppose the edges (v_1, v_2) and (v_3, v_4) are removed from Γ_D, separating the graph into two sub-cubic subgraphs $\Gamma_{S1} = \langle V_{S1}, E_{S1} \rangle$ and $\Gamma_{S2} = \langle V_{S2}, E_{S2} \rangle$. Without loss of generality, we assume that $v_1, v_3 \in V_{S1}$ and $v_2, v_4 \in V_{S2}$. Then, we define a cubic graph $\Gamma_1 = \langle V_1, E_1 \rangle$, where $V_1 = V_{S1}$ and $E_1 = E_{S1} \cup (v_1, v_3)$. Similarly, we define a second cubic graph $\Gamma_2 = \langle V_2, E_2 \rangle$, where $V_2 = V_{S2}$ and $E_2 = E_{S2} \cup (v_2, v_4)$. Then, Γ_D is obtained from the type 2 breeding operation $\mathscr{B}_2(\Gamma_1, \Gamma_2, v_1, v_3, v_2, v_4)$. This situation can be seen in Fig. 1.19.

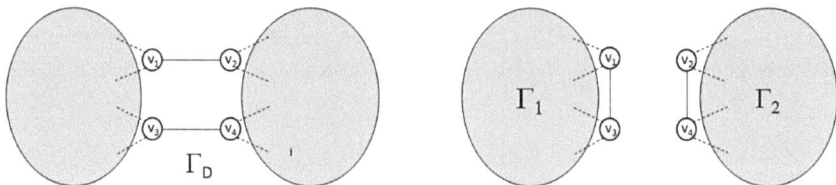

Fig. 1.19 Graphs Γ_D, Γ_1 and Γ_2 as described in Case 1 of Proposition 1.2

Case 2: At least one of the edges (v_1, v_3) or (v_2, v_4) is present in Γ_D. In this case, the 2-cracker cannot be obtained from a type 2 breeding operation, as such an operation would remove these edges. A 2-cracker of this type is therefore reducible. If both edges are present, we can focus on either one. Without loss of generality, we will assume that edge $(v_1, v_3) \in E_D$. Then, since Γ_D is cubic, and vertex v_1 is adjacent to vertices v_2 and v_3, it will be adjacent to one more vertex, say vertex a. Similarly, since vertex v_3 is adjacent to vertices v_1 and v_4, it will be adjacent to one more vertex, say vertex b. Note that it is possible that $a = b$, so we need to consider the cases separately.

Case 2.1: $a = b$. In this case, edge $(v_1, a) \in E_D$ and edge $(v_3, a) \in E_D$, as illustrated in Fig. 1.20. Since Γ_D is cubic, vertex a is adjacent to a third vertex, say c. Clearly, edge (a, c) is a bridge. Then, we define a cubic graph $\Gamma_1 = \langle V_1, E_1 \rangle$, where $V_1 = V_D \backslash \{v_1, v_3\}$ and $E_1 = E_D \backslash \{(v_1, v_2), (v_3, v_4), (v_1, v_3), (a, v_1), (a, v_3)\} \cup \{(a, v_2), (a, v_4)\}$. Then, we can obtain Γ_D from the type 3 parthenogenic operation $\mathscr{P}_3(\Gamma_1, a)$. Note that this case is essentially the same as Case 2.2 in Proposition 1.1. This situation can be seen in Fig. 1.20.

Case 2.2: $a \neq b$. It is obvious that the non-adjacent edges (v_1, a) and (v_3, b) form a cutset. This implies that either both edges are 1-crackers, or together they form a 2-cracker. The former situation is covered in Proposition 1.1 and Γ_D can be obtained from either type 1 breeding, type 1 parthenogenesis, or type 3 parthenogenesis. For the latter situation, let us define a cubic graph $\Gamma_1 = \langle V_1, E_1 \rangle$, where $V_1 = V_D \backslash \{v_1, v_3\}$ and $E_1 = E_D \backslash \{(v_1, v_2), (v_3, v_4), (v_1, v_3), (a, v_1), (b, v_3)\} \cup \{(a, v_2), (b, v_4)\}$, as illustrated in Fig. 1.21. We can then obtain Γ_D from the type 2 parthenogenic operation $\mathscr{P}_2(\Gamma_1, (a, v_2), (b, v_4))$. This situation can be seen in Fig. 1.21. $\square$

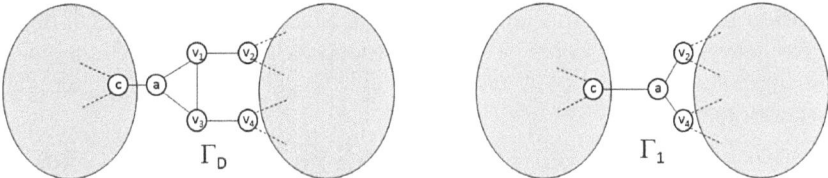

Fig. 1.20 Graphs Γ_D and Γ_1 as described in Case 2.1 of Proposition 1.2

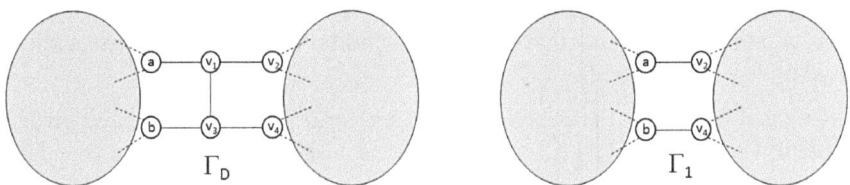

Fig. 1.21 Graphs Γ_D and Γ_1 as described in Case 2.2 of Proposition 1.2

Proposition 1.3. *Any descendant involving a 3-cracker can be obtained by performing type 3 breeding on appropriate subgraphs.*

Proof. Consider a descendant $\Gamma_D = \langle V_D, E_D \rangle$ containing a 3-cracker comprising edges (v_1, v_2), (v_3, v_4) and (v_5, v_6), as illustrated in Fig. 1.22. Then, suppose the edges (v_1, v_2), (v_3, v_4) and (v_5, v_6) are removed from Γ_D, separating the graph into two subgraphs $\Gamma_{S1} = \langle V_{S1}, E_{S1} \rangle$ and $\Gamma_{S2} = \langle V_{S2}, E_{S2} \rangle$. Without loss of generality, we assume that $\{v_1, v_3, v_5\} \in V_{S1}$ and $\{v_2, v_4, v_6\} \in V_{S1}$. Then, we introduce a new vertex v_7, and define a cubic graph $\Gamma_1 = \langle V_1, E_1 \rangle$, where $V_1 = V_{S1} \cup \{v_7\}$ and $E_1 = E_{S1} \cup \{(v_1, v_7), (v_3, v_7), (v_5, v_7)\}$. Similarly, we introduce a new vertex v_8, and define a second cubic graph $\Gamma_2 = \langle V_2, E_2 \rangle$, where $V_2 = V_{S2} \cup \{v_8\}$ and $E_2 = E_{S2} \cup \{(v_2, v_8), (v_4, v_8), (v_6, v_8)\}$. Then, we can obtain Γ_D from the type 3 breeding operation $\mathscr{B}_3(\Gamma_1, \Gamma_2, v_7, v_8, v_1, v_3, v_5, v_2, v_4, v_6)$. This situation can be seen in Fig. 1.22. $\square$

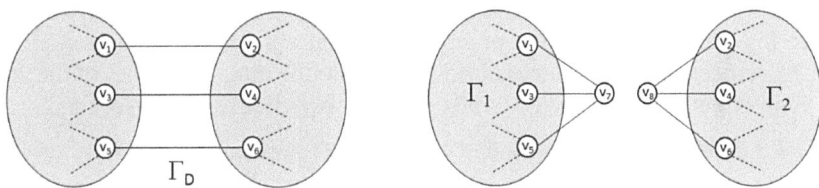

Fig. 1.22 Graphs Γ_D, Γ_1 and Γ_2 as described in Proposition 1.3

Note that any 3-cracker can be introduced via a type 3 breeding operation, and therefore all 3-crackers are irreducible.

Definition 1.11. We refer to a family of genes that can, through a series of breeding and parthenogenic operations, be used to produce a descendant Γ_D as a *complete family of ancestor genes* of Γ_D, and we say that Γ_D possesses this complete family of ancestor genes.

Definition 1.12. An *ancestral family tree* for a given descendant is a (directed) rooted tree where the root node corresponds to the descendant. Then, every node of the tree corresponds to an ancestor of the descendant, in such a way that node a is only connected to node b if the graph corresponding to node a is a parent of the graph corresponding to node b. Furthermore, we assume that any ancestral family tree is maximal, in the sense that all of the leaf nodes are genes (and so no additional parent nodes can be found).

Note that there may be several possible ancestral family trees for any given descendant. Propositions 1.1–1.3 and Definitions 1.11–1.12 allow us to propose the main theorem of this section.

Theorem 1.1. *Consider any descendant cubic graph Γ_D. Then,*

(1) Γ_D can be obtained from one or two parents by at least one of the six operations $\mathscr{B}_1(\cdot)$, $\mathscr{B}_2(\cdot)$, $\mathscr{B}_3(\cdot)$, $\mathscr{P}_1(\cdot)$, $\mathscr{P}_2(\cdot)$, $\mathscr{P}_3(\cdot)$, and

(2) Γ_D possesses a complete family of ancestor genes, which may be used to construct Γ_D.

Proof. Any descendant graph contains at least one cubic cracker. Any one of these cubic crackers can be selected, which will be either a 1-cracker, a 2-cracker, or a 3-cracker. Then, from Propositions 1.1–1.3, we can obtain Γ_D, from either one or two parents, using one of the six operations. Therefore (1) is proved.

Next, consider how Γ_D is obtained. If Γ_D is obtained through breeding, it has two parent graphs. If Γ_D is obtained through parthenogenesis, it has one parent graph. These parent graphs may be either genes, or descendants. If any of them are descendants, then by part (1) they also have one or two parent graphs each. Inductively, we can continue to consider the parents of descendants, while recording which operations are used to produce them, thereby obtaining an ancestral family tree of Γ_D. Once the entire tree is determined, all of the top nodes are genes, and we can recall the sequence of operations that produces Γ_D from these genes. Therefore (2) is proved.   □

See Fig. 1.23 for an example of an ancestral family tree for a descendant graph with 14 vertices. The three genes corresponding to the three leaf nodes of the ancestral family tree form a complete family of ancestor genes for the descendant.

Remark 1.2. The example illustrated in Fig. 1.23 shows that the family of ancestor genes of a cubic graph may contain multiple copies of the same gene. This is not unlike the presence of repeated eigenvalues in the unique spectrum of a square matrix. We have not emphasized this aspect in the present work as we focussed only on the decomposability of a graph into ancestral genes. However, in future investigations, the multiplicities of repeated ancestral genes may play a more significant role.

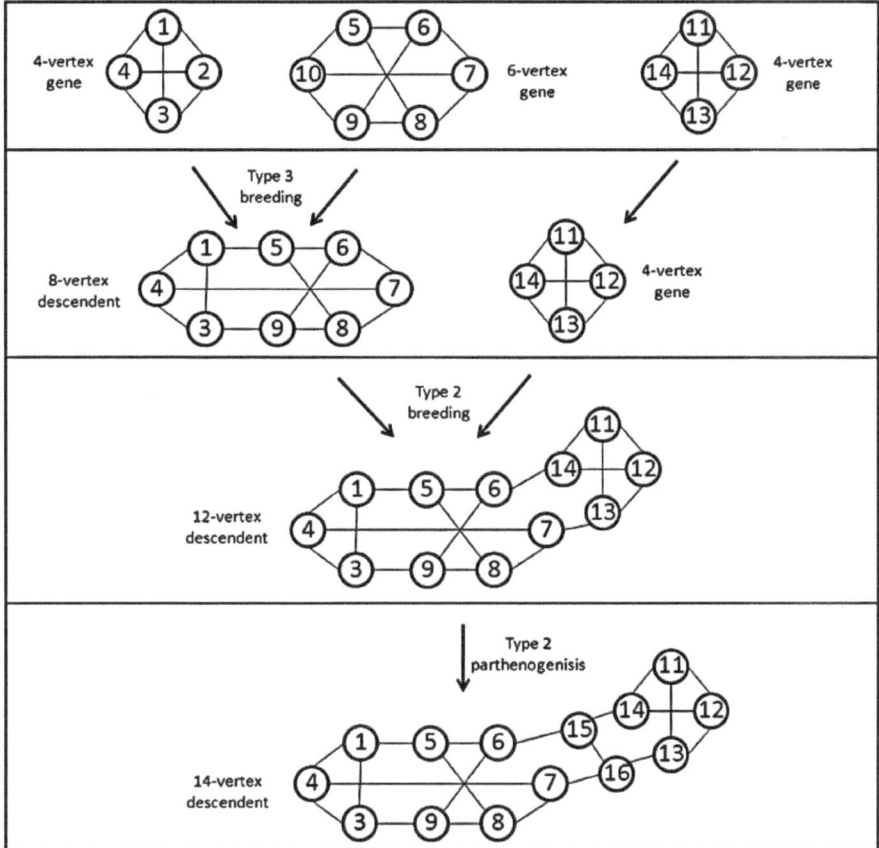

Fig. 1.23 An ancestral family tree for a 14-vertex descendant, with three ancestor genes

Theorem 1.1 indicates that, for any descendant, it is possible to obtain a complete family of ancestor genes by first applying an appropriate inverse operation to obtain one or two parents. Then we can apply an inverse operation on the parent(s) to obtain new parents (grandparents of the original descendant), and continue to apply inverse operations until a complete family of ancestor genes is obtained. However, it is conceivable that a descendant might be obtained for which no inverse operation is applicable. We will now prove that this situation cannot occur.

If a descendant contains any irreducible cubic crackers, an inverse breeding operation is possible. However, a given descendant may only contain reducible cubic crackers, meaning that there are no applicable inverse breeding operations. Therefore, in such a case, we can only possibly obtain a parent via an inverse parthenogenic operation, which produces the parent by removing a parthenogenic object from the descendant. In many cases, the removal of such a parthenogenic object changes a reducible cracker into an irreducible cracker which, in turn, 'permits an inverse breeding operation to be carried out. However, it is conceivable that we might encounter a descendant which contains neither irreducible cubic crackers, nor parthenogenic objects, in which case it would be impossible to recover any of this descendant's ancestors via any inverse operation. The following proposition ensures that such a situation never occurs, and furthermore that any reducible cracker in a given descendant can either be changed into an irreducible cracker, or removed entirely, by a sequence of inverse parthenogenic operations.

Proposition 1.4. *If a graph Γ_D is a descendant, one of the following must be true.*

(1) Γ_D contains at least one irreducible cubic cracker, or

(2) It is possible to perform a sequence of inverse parthenogenic operations, each time obtaining a new parent, until a parent is obtained that contains at least one irreducible cubic cracker.

Proof. Since Γ_D is a descendant, it contains one or more cubic crackers. If any of them are irreducible, then (1) is true. If all the cubic crackers are reducible, then Γ_D contains at least one 1-cracker or 2-cracker (since 3-crackers are always irreducible). We can therefore select either a reducible 1-cracker or a reducible 2-cracker in Γ_D. We will consider both cases separately.

Case 1: We select a reducible 1-cracker $C_1 = \{(a,b)\}$. Since Γ_D is cubic, we know that a must be adjacent to two more vertices, say c and d. Since C_1 is reducible, then without loss of generality, we can assume that edge (c,d) is present in Γ_D. Then, the cubicity of Γ_D also ensures that vertices c and d must each be adjacent to one more vertex, say e and f respectively. Note that it is possible that $e = f$.

Case 1.1: If $e = f$, then the cubicity of Γ_D ensures that vertex e must be adjacent to one more vertex, say g (see Fig. 1.24). Then, removing edges (a,b) and (e,g) from Γ_D disconnects a parthenogenic diamond, which we can remove from Γ_D by the use of the inverse type 1 parthenogenic operation $\mathscr{P}_1^{-1}(\Gamma_D, (a,b), (e,g))$. In the parent graph Γ_P, there is a new 1-cracker comprising edge (b,g). Note that the new 1-cracker may still be reducible.

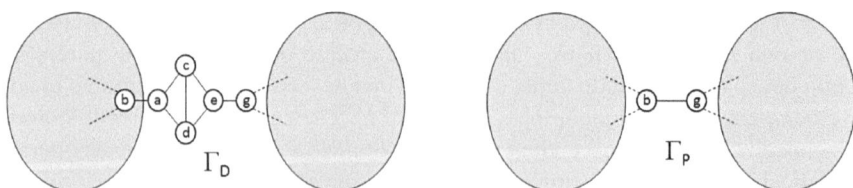

Fig. 1.24 Graphs Γ_D and Γ_P as described in Case 1.1

Case 1.2: If $e \neq f$, then removing edges (a,b), (c,e) and (d,f) from Γ_D isolates a parthenogenic triangle (see Fig. 1.25), which we can remove from Γ_D by the use of the inverse type 3 parthenogenic operation $\mathscr{P}_3^{-1}(\Gamma_D, a, c, d)$. In the parent graph Γ_P, the original 1-cracker $C_1 = \{(a,b)\}$ remains. Note that the 1-cracker may still be reducible.

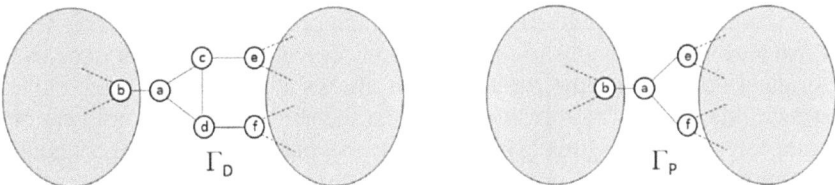

Fig. 1.25 Graphs Γ_D and Γ_P as described in Case 1.2

Case 2: We select a reducible 2-cracker $C_2 = \{(a,b),(c,d)\}$. Since C_2 is reducible, then without loss of generality, we can assume that edge (a,c) is present in Γ_D. Then, the cubicity of Γ_D also ensures that vertices a and c must each be adjacent to one more vertex, say e and f, respectively. Note that it is possible that $e = f$.

Case 2.1: If $e = f$, then as illustrated in Fig. 1.26, the cubicity of Γ_D ensures that vertex e must be adjacent to one more vertex, say g. Then, removing edges (a,b), (c,d) and (e,g) from Γ_D disconnects a parthenogenic triangle, which we can remove from Γ_D by the use of the inverse type 1 parthenogenic operation $\mathscr{P}_1^{-1}(\Gamma_D,a,c)$. In the parent graph Γ_P, a 1-cracker comprising edge (e,g) remains. Note that this 1-cracker may still be reducible.

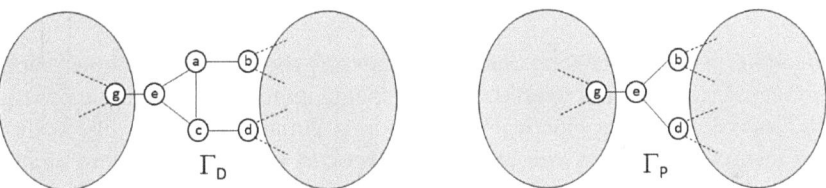

Fig. 1.26 Graphs Γ_D and Γ_P as described in Case 2.1

Case 2.2: If $e \neq f$, then as illustrated in Fig. 1.27, the removal of edges (a,b), (c,d), (a,e) and (c,f) from Γ_D disconnects a parthenogenic bridge, which we can remove from Γ_D by the use of the inverse type 2 parthenogenic operation $\mathscr{P}_2^{-1}(\Gamma_D,a,c)$. In the parent graph Γ_P, there is a new 2-cracker comprising edges (e,b) and (f,d). Note that this 2-cracker may still be reducible.

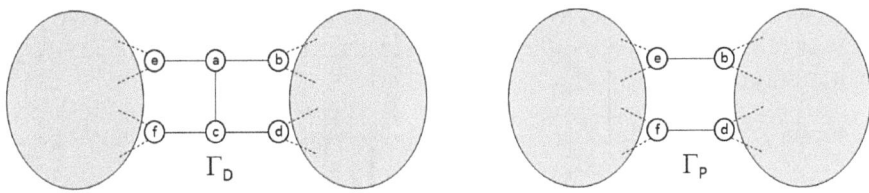

Fig. 1.27 Graphs Γ_D and Γ_P as described in Case 2.2

Since in all cases considered, the parent graph contains a cubic cracker (either a new one introduced by an inverse parthenogenic operation, or one that remains from the original descendant), the parent is itself always a descendant. Then, either (1) is true for this parent, or if not, we can repeat the above procedure until we obtain a parent for which (1) is true. Since each inverse parthenogenic process outputs a parent with fewer vertices than its child, this process is guaranteed to eventually converge to such a case. □

1.6 Conclusions

The theory presented above allows us to partition the set S of connected cubic graphs into two distinct and exhaustive categories – the comparatively smaller set of genes, which form the basic building blocks of S, and the much larger set of descendants, which inherit a lot of their structure from the genes. Theorem 1.1 and Proposition 1.4 give both a proof of existence, and a constructive method to obtain a complete family of ancestor genes for any given descendant. An algorithm to identify a complete family of ancestor genes, given a descendant, would be a simple task of identifying all of the cubic crackers in the descendant, examining each until one is found that permits an inverse operation, and recursively repeating the process in each parent obtained until only genes remain. Such an algorithm would terminate in polynomial time.

In addition to the aesthetic beauty of rendering the ancestry of cubic graphs as finite sets of smaller cubic graphs, there are some important algorithmic benefits as well. Since descendants inherit much of their structure from genes, the search for graph theoretic properties within a descendant can sometimes be more efficiently recast as a search of the (potentially much smaller) genes instead. In this context, Theorem 1.1 and Proposition 1.4 give rise to a generic decomposition algorithm that could be utilised for any descendant. Since the ancestor genes also lie within the set of connected cubic graphs, any existing algorithms designed for cubic graphs will be applicable for the genes as well. An initial exploration of the concept of inheritance from ancestor genes follows in Chap. 2.

Given that properties can be inherited from a complete family of ancestor genes, a natural question to ask is how many different complete families of ancestor genes a descendant graph might have. If it were possible for a descendant to be constructed from several different complete families of ancestor genes, it could be the case that the different complete families might have different properties, which would make analysis of their inheritance difficult. However, the following theorem removes any such confusion.

Theorem 1.2. *Any descendant Γ_D has a unique complete family of ancestor genes.*

Theorem 1.2 provides, along with Theorem 1.1, a guarantee of existence and uniqueness of a complete family of ancestor genes for every cubic graph. For such a graph, Theorem 1.2 implies that the graph is either a gene, or there is a unique complete family of ancestor genes which can be identified in polynomial time, and that it is this particular complete family of ancestors genes which provides the majority of structure within the descendant. It should be noted that although Theorem 1.2 ensures that a given descendant may only be constructed from a unique complete family of ancestor genes, it does not guarantee that only one set of breeding operations may be used.

The full proof of Theorem 1.2 is rather long and technical, and so we leave it for Chap. 3. However, we use the result as motivation for the work presented in Chap. 2.

Remark 1.3. We conclude this chapter with two examples of countably infinite sequences of genes. As mentioned earlier, any non-trivial Snark is a gene, so the famous Flower Snarks [13] are one example of a countably infinite sequence of non-Hamiltonian genes.

A second countably infinite sequence of genes, which are all Hamiltonian, is the family of Möbius ladder graphs [10], which can be constructed for any size $2m \geq 6$. Their construction is very simple; start with a cycle graph of size $2m$, and then add edges $(i, m+i)$ for all $i = 1, \ldots, m$. An example of a Möbius ladder graph of size 16 is displayed in Fig. 1.28. Clearly this graph is cubic and Hamiltonian. To see that it is a gene, the following argument can be used.

Suppose a cubic cracker C exists in a Möbius ladder graph of size $2m$. It is clear that the removal of any number of the edges $(i, m+i)$ alone cannot disconnect any vertices, so C must contain at least two (non-adjacent) edges from the cycle. Suppose C contains exactly two edges from the cycle. Their removal separates the cycle into two segments, however there will always be at least two edges of the form $(i, m+i)$

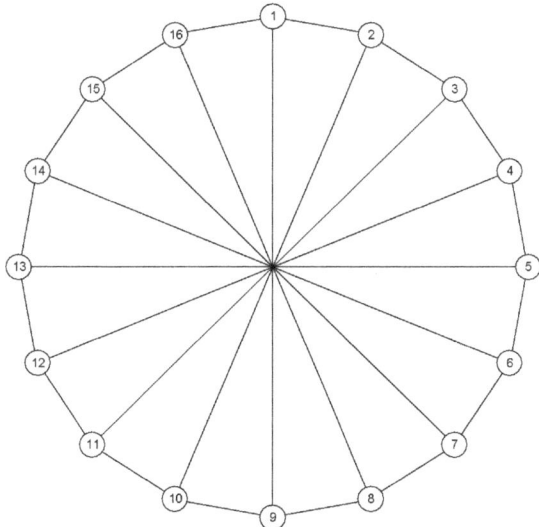

Fig. 1.28 A Möbius ladder graph of size 16

between the two segments, and so C cannot be a cubic cracker, violating the initial assumption. Therefore C must contain exactly three edges from the cycle. However, by same argument as above, removing these three edges separates the cycle into three segments, and emanating from each segment will be at least two edges of the form $(i, m+i)$. Then the removal of these three edges does not disconnect the graph, which is a contradiction. So Möbius ladder graphs do not contain any cubic crackers and, therefore, constitute genes.

Note that all Möbius ladder graphs of size $2m \geq 8$ or larger contain exactly m 4-cycles of the form $i \to i+m \to i+m+1 \mod 2m \to i+1 \to i$ for any $i = 1, \ldots, m$. Therefore, since they contain no cubic crackers, from Lemma 1.2 we see that their smallest cracker must be of size 4. The Möbius ladder of size 6 is the graph Γ_6^* discussed in Sect. 1.3 which contains no cubic crackers.

References

1. Balaban, A.T., Harary, F.: The characteristic polynomial does not uniquely determine the topology of a molecule. J. Chem. Doc. **11**, 258–259 (1971)
2. Barnette, D.: Recent progress in combinatorics. In: Tutte, W.T. (ed.) Proceedings of the Third Waterloo Conference on Combinatorics, May 1968. Academic, New York (1969)
3. Batagelj, V.: Inductive classes of graphs. In: Proceedings of the Sixth Yugoslav Seminar on Graph Theory, Dubrovnik, 1985, pp. 43–56 (1986)
4. Blanuša, D.: Problem cetiriju boja. Glasnik Mat. Fiz. Astr. Ser. II **1**, 31–42 (1946)
5. Brinkmann, G.: Fast generation of cubic graphs. J. Graph Theory **23**, 139–149 (1996)
6. Brinkmann, G., Goedgebeur, J., McKay, B.D.: Generation of cubic graphs. Discret. Math. Theor. Comput. Sci. **13**(2), 69–80 (2011)
7. Ding, G., Kanno, J.: Splitter theorems from cubic graphs. Comb. Probab. Comput. **15**, 355–375 (2006)
8. Eppstein, D.: The traveling salesman problem for cubic graphs. In: Dehne, F., Sack, J-R., Smid, M. (eds.) Algorithms and Data Structures. Volume 2748 of Lecture Notes in Computer Science, pp. 307–318. Springer, Heidelberg (2003)
9. Filar, J.A., Haythorpe, M., Nguyen, G.T.: A conjecture on the prevalence of cubic bridge graphs. Discuss. Math. Graph Theory **30**(1), 175–179 (2010)
10. Guy, R.K., Harary, F.: On the Möbius ladders. Can. Math. Bull. **10**, 493–496 (1967)
11. Harary, F.: Graph Theory. Addison-Wesley, Reading (1969)
12. Holton, D.A., Sheehan, J.: The Petersen Graph. Cambridge University Press, Cambridge (1993)
13. Isaacs, R.: Infinite Families of Nontrivial Trivalent Graphs Which Are Not Tait Colorable. Am. Math. Mon. **82**, 221–239 (1975)
14. Lou, D., Teng, L., Wu, X.: A polynomial algorithm for cyclic edge connectivity of cubic graphs. Australas. J. Comb. **24**, 247–259 (2001)
15. McKay, B.D., Royle, G.F.: Constructing the cubic graphs on up to 20 vertices. Ars Comb. **12A**, 129–140 (1986)
16. Meringer, M.: Fast generation of regular graphs and construction of cages. J. Graph Theory **30**, 137–146 (1999)
17. Ore, O.: The Four-Colour Problem. Academic, New York (1967)
18. Read, R.C., Wilson, R.J.: An Atlas of Graphs, p. 263. Oxford University Press, Oxford (1998)
19. Royle, G.: Constructive enumeration of graphs. Ph.D. thesis, University of Western Australia (1987)

Chapter 2
Inherited Properties of Descendants

Abstract Given the results of Chap. 1 that every descendant may be constructed from a complete family of ancestor genes, we now investigate how the properties of the descendant correspond to the properties of those genes. In particular we consider the properties of Hamiltonicity, bipartiteness and planarity. In all three cases we prove that a descendant may only possess the property if all of its ancestor genes do. In the case of bipartiteness and planarity we also establish sufficient conditions. These results allow us to analyse the properties of a graph by considering its ancestor genes, or alternatively, to construct a graph with desired properties by choosing smaller genes with those properties. We follow each section with a discussion of famous results and conjectures relating to the graph properties, and how the results of this chapter relate to them.

2.1 Introduction

The results of the previous chapter indicate that every cubic graph is either a gene, or can be constructed from a unique complete family of ancestor genes via some sequence of the six breeding operations. It is therefore obvious that any such descendant has obtained the entirety of its structure either from its complete family of ancestor genes, or via the breeding operations used to construct it. A reasonable assumption is that some of the graph theoretic properties present in the descendant may have been inherited from its ancestor genes. In this chapter we investigate three such properties – Hamiltonicity, bipartiteness and planarity – and demonstrate that this is indeed the case. Specifically, we will prove that, in order for a descendant to possess any of the three properties, it is necessary for all of its ancestor genes to possess the property as well. In the cases of bipartiteness and planarity, this condition is both necessary and sufficient. This chapter not intended to represent an exhaustive list of inherited properties in descendants, but merely a first exploration into such inheritance. The discovery of other such inherited properties is a ripe subject for future research.

© Springer International Publishing Switzerland 2016

P. Baniasadi et al., *Genetic Theory for Cubic Graphs*, SpringerBriefs in Operations Research, DOI 10.1007/978-3-319-19680-0_2

2.2 Hamiltonicity

The *Hamiltonian cycle problem* is a famously difficult graph theory problem that is known to be NP-complete, even when restricted to the class of cubic graphs [7].[1] The problem can be stated simply: given a graph Γ containing N vertices, determine whether or not any simple cycles of length N exist in the graph. Such a simple cycle of length N is called a *Hamiltonian cycle*. If a graph contains at least one Hamiltonian cycle, it is said to be a *Hamiltonian* graph, whereas a graph with no Hamiltonian cycles is said to be a *non-Hamiltonian* graph. A Hamiltonian graph is said to possess the property of *Hamiltonicity*.

Rather than focusing on Hamiltonian descendants, in this section we primarily investigate the different methods in which a non-Hamiltonian descendant may be constructed, as non-Hamiltonian cubic graphs are known to be relatively rare (e.g. see Robinson and Wormald [15]) and are therefore simpler to characterise. Recall from Chap. 1 that non-Hamiltonian genes are called *mutants*. We separate non-Hamiltonian descendants into two categories – mutant descendants, which possess at least one mutant in their complete family of ancestor genes; and non-mutant non-Hamiltonian (NMNH) descendants. In order to aid the flow of proofs, we consider the latter category first.

2.2.1 Non-mutant Non-Hamiltonian Descendants

Consider a descendant graph Γ_D. From Theorem 1.2, we know that Γ_D has a unique complete family of ancestor genes. Suppose that none of these genes are mutants. Then in general Γ_D could be Hamiltonian or non-Hamiltonian. We now investigate, exhaustively, all methods by which non-Hamiltonian descendants can be constructed from Hamiltonian parents for each of the six breeding operations.

2.2.1.1 Type 1 Breeding

Lemma 2.1. *If a descendant graph Γ_D is output by the type 1 breeding operation $\mathscr{B}_1(\Gamma_1, \Gamma_2, e_1, e_2)$ for some Γ_1, Γ_2, e_1 and e_2, then Γ_D is non-Hamiltonian (Fig. 2.1).*

Proof. Type 1 breeding always introduces a 1-cracker into the graph, and so Γ_D is a bridge graph and therefore non-Hamiltonian. □

Corollary 2.1. *Any descendants of the descendant Γ_D described in Lemma 2.1 are also non-Hamiltonian.*

---

[1] In fact, the Hamiltonian cycle problem is still NP-complete even for cubic planar graphs [8]. Cubic planar graphs are investigated in Sect. 2.4 of the present manuscript.

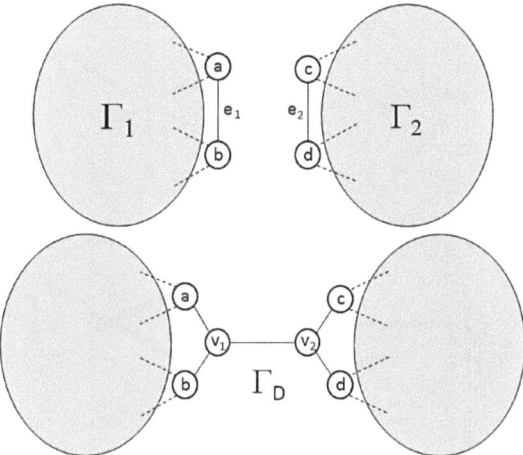

Fig. 2.1 A descendant obtained from type 1 breeding, and its two parents

Proof. Γ_D was produced by a type 1 breeding operation, and therefore contains a 1-cracker. This 1-cracker is preserved regardless of future operations, because type 2 and 3 breeding operations are not permitted where they would break the 1-cracker, type 2 parthenogenesis is only permitted on 2-crackers, and types 1 and 3 parthenogenesis preserve (by the introduction of equivalent) 1-crackers. Therefore, any descendants of Γ_D also contain at least one 1-cracker and are themselves bridge graphs. □

Type 1 breeding is the easiest method of producing NMNH descendants, as it can be performed on any two parents, and is guaranteed to output a NMNH descendant, specifically a bridge graph.

2.2.1.2 Type 2 Breeding

Recall that when type 2 breeding is performed, an edge is broken in each of two parent graphs, and the broken edges are joined together to form the descendant. This process always outputs a 2-connected graph, and so any Hamiltonian cycle in the descendant must traverse both edges in the 2-cracker introduced by the type 2 breeding operation.

Definition 2.1. If a graph Γ contains an edge e that is never traversed in any Hamiltonian cycles in Γ, we say that e is a *non-Hamiltonian edge* (NH-edge).

Note that, by definition, every edge in a non-Hamiltonian graph is an NH-edge.

Lemma 2.2. *Consider a descendant graph Γ_D formed by the type 2 breeding operation $\mathscr{B}_2(\Gamma_1, \Gamma_2, a, b, c, d)$, for some $\Gamma_1 = \langle V_1, E_1 \rangle$, $\Gamma_2 = \langle V_2, E_2 \rangle$, $e_1 := (a,b) \in E_1$ and $e_2 := (c,d) \in E_2$. If either e_1 or e_2 is a non-Hamiltonian edge, then Γ_D is non-Hamiltonian.*

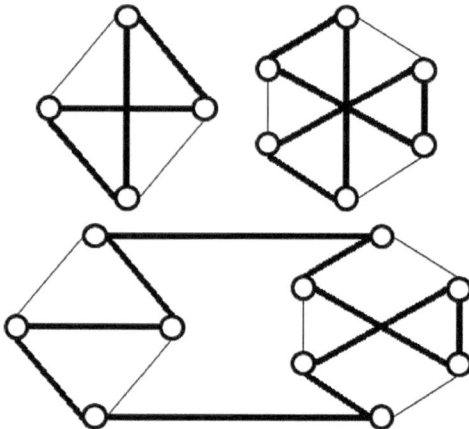

Fig. 2.2 Hamiltonian cycles in two parents, and the corresponding Hamiltonian cycle in one of their descendants

Proof. Any Hamiltonian cycle in Γ_D must be composed from a Hamiltonian cycle in Γ_1 that traverses e_1, and a Hamiltonian cycle in Γ_2 that traverses e_2 (e.g. see Fig. 2.2). However, at least one of e_1 and e_2 is a non-Hamiltonian edge, and therefore these two Hamiltonian cycles cannot both exist. Therefore, Γ_D has no Hamiltonian cycles. □

Corollary 2.2. *Without loss of generality, if Γ_1 is non-Hamiltonian, then Γ_D is non-Hamiltonian as well.*

Proof. Since Γ_1 is non-Hamiltonian, it is clear that e_1 is an NH-edge. Then the result follows immediately from Lemma 2.2. □

NH-edges can occur naturally in Hamiltonian genes. The smallest such example, on 14 vertices, is displayed in Fig. 2.3. However, NH-edges are more commonly induced by selective breeding, such as in the following example.

Lemma 2.3. *Consider a descendant Γ_D formed by the type 2 parthenogenic operation $\mathscr{P}_2(\Gamma_1, e_1, e_2)$. Then, the parthenogenic bridge (v_1, v_2) in Γ_D constitutes an NH-edge.*

Proof. Consider a Hamiltonian cycle in Γ_D that traverses the parthenogenic bridge. For convenience, assume the Hamiltonian cycle begins on vertex v_2. Then, once the parthenogenic bridge is crossed, there are two choices of which vertex to visit next, say vertex a or vertex b. If we visit vertex a, we can no longer visit vertex b without revisiting a vertex, and vice versa. This is because it is now impossible to get to the other side of the parthenogenic bridge without closing the cycle. This situation can be seen in Fig. 2.4. Therefore, no Hamiltonian cycles can use the parthenogenic bridge, and it constitutes an NH-edge. □

Another method of inducing an NH-edge is to manually select a set of edges that do not exist simultaneously in any Hamiltonian cycle in a graph.

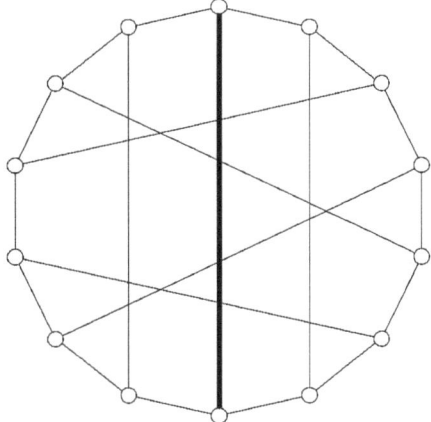

Fig. 2.3 The smallest Hamiltonian gene containing an NH-edge, which is shown here in *bold*

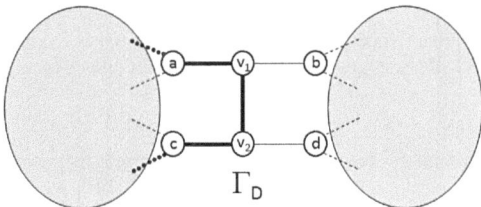

Fig. 2.4 A (short) cycle using a parthenogenic bridge

Definition 2.2. An *NH-edge-set* for a graph is a set of edges that are not all traversed in any single Hamiltonian cycle in that graph.

For example, any three edges adjacent to a single vertex in a cubic graph form an NH-edge-set of cardinality three. It is clear that after performing type 2 breeding on all but one edge in an NH-edge-set, the remaining edge constitutes an NH-edge. An example of a NHNM descendant produced using this method is shown in Fig. 2.5. Of course, an NH-edge is merely a special case of an NH-edge-set, with cardinality 1.

2.2.1.3 Type 3 Breeding

Recall that when type 3 breeding is performed, a vertex is removed from each parent graph, and the three vertices from each parent which were adjacent to the removed vertices are joined together by a 3-cracker. Since this process provides three passages between the two sides of the 3-cracker, any Hamiltonian cycle in the descendant must traverse exactly 2 of the 3 edges. It is important to note that any Hamiltonian cycle must travel through all vertices on one side of the 3-cracker first before crossing the 3-cracker.

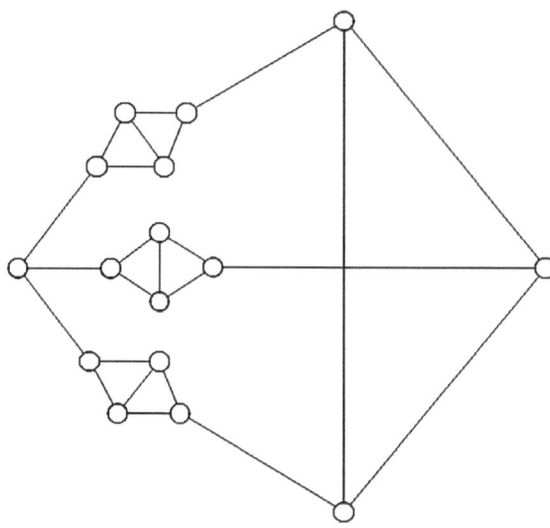

Fig. 2.5 A non-Hamiltonian graph formed by performing type 2 breeding on all edges in a NH-edge-set consisting of three edges adjacent to a vertex in one of the parents

Consider a descendant Γ_D formed by the type 3 breeding operation $\mathscr{B}_3(\Gamma_1, \Gamma_2, v_1, v_2, a, b, c, d, e, f)$, for some $\Gamma_1 = \langle V_1, E_1 \rangle$, $\Gamma_2 = \langle V_2, E_2 \rangle$, $v_1, a, b, c \in V_1$ and $v_2, d, e, f \in V_2$. Then if a Hamiltonian cycle in Γ_D traverses edges (a, d) and (b, e) say, it can be composed from two smaller Hamiltonian cycles, one in each parent. The corresponding Hamiltonian cycle in Γ_1 must contain the path $a - v_1 - b$, and the corresponding Hamiltonian cycle in Γ_2 must contain the path $d - v_2 - e$. This situation is displayed in Fig. 2.6.

Note that the presence of an NH-edge in a parent is not sufficient on its own to permit the construction of a non-Hamiltonian descendant via type 3 breeding, because the other two edges adjacent to the same vertex can be used instead. Also, it is clear that in a Hamiltonian graph, no two NH-edges can be adjacent. Therefore, producing a non-Hamiltonian descendant from Hamiltonian parents with type 3 breeding requires the consideration of *forced edges*, which are edges that exist in every Hamiltonian cycle in a given graph.

Lemma 2.4. *If a cubic Hamiltonian graph contains an NH-edge e, then the four adjacent edges to e are all forced edges.*

Proof. Hamiltonian cycles traverse two edges adjacent to each vertex. Edge e is adjacent to two vertices, and so the two remaining adjacent edges for both vertices must be used in every Hamiltonian cycle. □

Lemma 2.5. *Consider a descendant graph Γ_D formed by the type 3 breeding operation $\mathscr{B}_3(\Gamma_1, \Gamma_2, v_1, v_2, a, b, c, d, e, f)$, for some $\Gamma_1 = \langle V_1, E_1 \rangle$, $\Gamma_2 = \langle V_2, E_2 \rangle$, $v_1, a, b, c \in V_1$ and $v_2, d, e, f \in V_2$. Without loss of generality, if $(a, v_1) \in E_1$ is a forced edge, and $(d, v_2) \in E_2$ is an NH-edge, then Γ_D is non-Hamiltonian.*

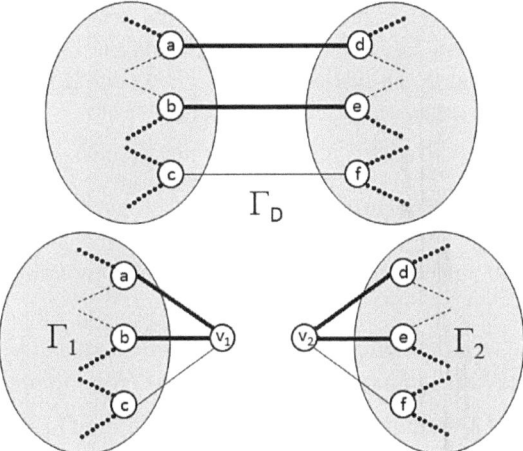

Fig. 2.6 A Hamiltonian cycle in a descendant obtained from type 3 breeding, and the corresponding Hamiltonian cycles in the parents

Proof. Any Hamiltonian cycle in Γ_D must be composed of a Hamiltonian cycles in each of Γ_1 and Γ_2. Since (a, v_1) is a forced edge, it must be traversed in all Hamiltonian cycles in Γ_1, and therefore 3-cracker edge (a, d) must be traversed in any corresponding Hamiltonian cycle in Γ_D. However, this implies that edge (d, v_2) must be traversed in any corresponding Hamiltonian cycle in Γ_2. Edge (d, v_2) is an NH-edge, and so no Hamiltonian cycle traversing this edge can be found in Γ_2. Therefore, there are no Hamiltonian cycles in Γ_D. □

Corollary 2.3. *Without loss of generality, if Γ_1 is non-Hamiltonian, then Γ_D is also non-Hamiltonian.*

Proof. In Lemma 2.5, non-Hamiltonicity results because it is impossible to avoid traversing an NH-edge. If Γ_1 is non-Hamiltonian, all edges $e \in E_1$ are NH-edges, and it is again impossible to avoid traversing one. Then, by the same arguments used in the proof of Lemma 2.5, Γ_D is non-Hamiltonian. □

Lemma 2.6. *If Γ_1 and Γ_2 are Hamiltonian, and it is not the case that one graph contains an NH-edge and the other contains a forced edge, then any type 3 breeding operation involving Γ_1 and Γ_2 outputs a Hamiltonian graph.*

Proof. If neither graph contains an NH-edge then the result is trivial. Consider the situation where a graph contains NH-edges. As stated earlier, no two NH-edges can be adjacent in a Hamiltonian cubic graph, and therefore any 3-cracker involving an NH-edge is still traversable by a Hamiltonian cycle by using the other two edges. Only if the NH-edge is demanded does non-Hamiltonicity result, which requires the presence of a forced edge in the second parent. □

While forced edges do occur naturally in some genes, inducing them in descendants is a simple task; any type 2 breeding operation produces a descendant containing a 2-cracker, and if the descendant is Hamiltonian, this 2-cracker comprises two forced edges. Inducing NH-edges also results in the production of forced edges, as demonstrated by Lemma 2.4. It should be noted that performing a type 3 breeding operation cannot introduce an NH-edge or a forced edge if the two parents did not contain any to begin with.

2.2.1.4 Parthenogenic Operations

It is easy to see that parthenogenic operations alone cannot alter the Hamiltonicity of a graph.

Lemma 2.7. *If Γ_D is a descendant produced by any parthenogenic operation, and Γ_1 is its parent graph, then Γ_D and Γ_1 have the same Hamiltonicity.*

Proof. If types 1 or 3 parthenogenesis are used to create Γ_D, then by definition, Γ_1 must be a bridge graph, and the result follows immediately from Corollary 2.1. So we only need to consider type 2 parthenogenesis. Then, Γ_1 contains a 2-cracker, comprising edges, say, (a,b) and (c,d).

Consider first the case where Γ_1 is Hamiltonian. Clearly, both edges in the 2-cracker are forced edges. Consider the effect of placing a degree-2 vertex inside each of these edges, say vertices v_1 and v_2. That is, edge (a,b) is replaced with edges (a,v_1), (v_1,b), and edge (c,d) is replaced with edges (c,v_2) and (v_2,d). It is clear that the resulting graph is still Hamiltonian, and that every Hamiltonian cycle in Γ_1 corresponds to a Hamiltonian in the new graph with the 2-cracker edges replaced accordingly. Then, the addition of the edge (v_1,v_2), which gives Γ_D, cannot destroy Hamiltonicity, and therefore Γ_D is Hamiltonian as well (Fig. 2.7).

Next, consider the case where Γ_1 is non-Hamiltonian. Similar to the argument in the previous case, we can introduce vertices v_1 and v_2 inside the 2-cracker edges without altering the Hamiltonicity. Then, the addition of the edge (v_1,v_2), which gives Γ_D, also retains non-Hamiltonicity, because from Lemma 2.3, edge (v_1,v_2) is an NH-edge. Therefore Γ_D is non-Hamiltonian. □

2.2.2 Main Theorem and Discussion

The results so far allow us to propose the main theorem of this section.

Theorem 2.1. *If a descendant has at least one non-Hamiltonian parent, then the descendant is also non-Hamiltonian.*

Proof. The descendant can only be obtained from one of the three breeding operations, or one of the three parthenogenic operations. Then, the result follows immediately from Corollaries 2.1–2.3 and Lemma 2.7. □

The strength of Theorem 2.1 is that it ensures that once a non-Hamiltonian descendant has been constructed, it is impossible to "reintroduce" Hamiltonicity in future descendants.

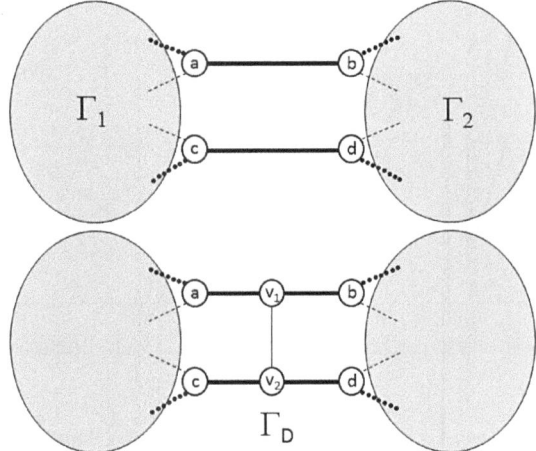

Fig. 2.7 A Hamiltonian cycle in a descendant involving a 2-cracker, and the corresponding cycle in its child containing a parthenogenic bridge

2.2.2.1 Mutant Descendants

Consider a descendant graph Γ_D. From Theorem 1.2, we know that Γ_D has a unique complete family of ancestor genes. Suppose that one of these genes is a mutant. We then refer to Γ_D as a *mutant descendant*.

Lemma 2.8. *If a descendant Γ_D is a mutant descendent, then it is non-Hamiltonian.*

Proof. The proof follows immediately from Theorems 2.1 and 1.2. □

Figure 2.8 shows Tietze's graph, which is the smallest mutant descendant.

2.2.2.2 Relative Cardinalities of Non-Hamiltonian Graphs

We could now partition non-Hamiltonian cubic graphs into three disjoint sets: mutant descendants, NMNH descendants and mutants. However, note that bridge graphs inhabit both of the former two sets. For the sake of simplicity, we will treat bridge graphs as their own category of non-Hamiltonian graphs, and thus we partition non-Hamiltonian cubic graphs into four disjoint sets: cubic bridge graphs, (non-bridge) mutant descendants, (non-bridge) NMNH descendants, and mutants. By convention, when we refer to mutant descendants or NMNH descendants, it will be assumed that such descendants are not bridge graphs. We refer to the set of graphs of size N in each of these four categories as $\mathcal{N}\mathcal{H}_N^b$, $\mathcal{N}\mathcal{H}_N^{md}$, $\mathcal{N}\mathcal{H}_N^{nm}$ and $\mathcal{N}\mathcal{H}_N^m$ respectively.

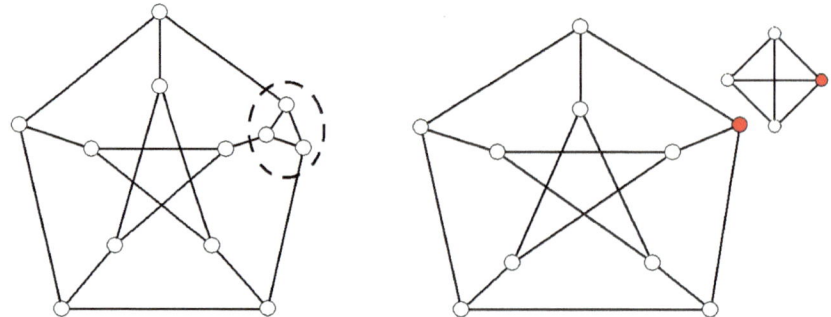

Fig. 2.8 Tietze's graph, and its parents, which include the (mutant) Petersen graph

In [6] it was conjectured that the relative cardinality of cubic bridge graphs of a given size, as a ratio of all non-Hamiltonian graphs of the same size, converges to 1 as the size increases. We display in Table 2.1 the relative cardinalities of the four different types of non-Hamiltonian cubic graphs for different fixed sizes.

N	$\mathscr{N}\mathscr{H}_N^b$ (%)	$\mathscr{N}\mathscr{H}_N^{md}$ (%)	$\mathscr{N}\mathscr{H}_N^{nm}$ (%)	$\mathscr{N}\mathscr{H}_N^m$ (%)	# of NH graphs
10	50.00	0	0	50.00	2
12	80.00	20.00	0	0	5
14	82.86	14.29	2.86	0	35
16	84.93	11.87	3.20	0	219
18	86.13	11.28	2.46	0.12	1666
20	87.40	9.53	3.02	0.05	14498
22	88.59	8.65	2.74	0.02	148790

Table 2.1 Relative cardinalities of the four different types of non-Hamiltonian cubic graphs containing up to 20 vertices

Conjecture 2.1. Consider the set $\mathscr{N}\mathscr{H}_N^{nb}$ of all non-bridge non-Hamiltonian cubic graphs of size N. In terms of the categories given in Table 2.1, we have $\mathscr{N}\mathscr{H}_N^{nb} = \mathscr{N}\mathscr{H}_N^{md} \cup \mathscr{N}\mathscr{H}_N^{nm} \cup \mathscr{N}\mathscr{H}_N^m$. Then,

$$\lim_{N \to \infty} \frac{\mathscr{N}\mathscr{H}_N^{nm}}{\mathscr{N}\mathscr{H}_N^{nb}} = 1.$$

That is, NMNH descendants become the dominant set of non-bridge non-Hamiltonian graphs as the size of the graph increases. This conjecture is natural extension of the conjecture given in [6]. Bridge graphs are relatively easy to produce, as they can be constructed from any two parents through the use of type 1 breeding. Similarly, NMNH descendants can be constructed from any three parents, through the use of type 2 breeding, type 2 parthenogenesis to create a parthenogenic bridge (which constitutes an NH-edge by Lemma 2.3), and then type 2 breeding on the parthenogenic bridge. Mutant descendants, on the other hand, can only be constructed from the extremely rare mutants. It is this line of thought that motivates the conjecture. It should be noted that in Table 2.1 the proportion of mutant descendants is still larger than the proportion of NHNB descendants for $N = 22$. However, that proportion appears to be steadily decreasing, while the proportion of NMNH descendants is staying roughly constant. It will be necessary to consider larger sets of non-Hamiltonian graphs to gain a clearer picture.

2.3 Bicubic Graphs

A bipartite graph is one whose vertices can be divided into two disjoint subsets, such that every edge in the graph is adjacent to exactly one vertex from each subset. Alternatively, we can think of bipartite graphs as being those which permit each vertex to be assigned a binary weight (either $+1$ or -1), such that the weight of every edge (equal to the sum of weights on its two incident vertices) in the graph is zero. Bipartite graphs are a commonly studied class of graphs in literature. In particular, bipartite cubic graphs (also known as bicubic graphs) were the subject of Tutte's Conjecture that all 3-connected bicubic graphs are Hamiltonian, for which a counterexample was first discovered by Horton [3]. The Horton graph is discussed in some detail at the end of this section. A refinement of Tutte's conjecture, with the additional requirement of planarity, was suggested by Barnette [2] and remains an open problem. Planarity is discussed in detail in Sect. 2.4 of the present manuscript.

In this section, we consider the construction of bicubic descendants. Certainly there exist bicubic genes – the 6-vertex gene Γ_6^* is the smallest such example. Bicubic mutants also exist (and constitute counterexamples to Tutte's Conjecture), with George's graph the smallest currently known, on 50 vertices [9]. Whether any or not any smaller bicubic mutants exist is an open problem. The 6-vertex gene and George's graph are displayed in Fig. 2.9.

To prove that a graph is bipartite, we need to demonstrate that we can assign weights of $+1$ or -1 to each vertex in such a way that all edge weights are zero. For cubic graphs (and in fact any k-regular graphs for odd k), this determination can sometimes be aided by the following result.

Lemma 2.9. *Consider a cubic graph* $\Gamma = \langle V, E \rangle$. *It is impossible to assign weights of* $+1$ *or* -1 *to each vertex in such a way that all but one edge has zero weight.*

Proof. The weight on each edge $e \in E$ is equal to the sum of the weights on its two adjacent vertices. That is, if w_i is the weight assigned to vertex $i \in V$, then the weight of an edge $(i, j) \in E$ is $w_i + w_j$. Note that, as vertices can only be assigned a weight of $+1$ or -1, an edge may only have a weight of $+2$, 0, or -2. Since Γ is cubic, each vertex is adjacent to three edges. Therefore, the total sum W of edge weights in Γ is

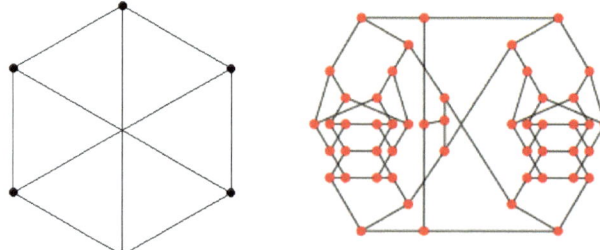

Fig. 2.9 The two smallest known Hamiltonian and non-Hamiltonian bicubic genes, Γ_6^* and George's graph respectively

$$W = 3\sum_{i=1}^{|V|} w_i.$$

Since all of the vertex weights are integer, it is clear that W must be some factor of 3. However, if we require all but one edge in E to have zero weight, then W must be either 2 or -2. Neither value is a factor of 3, and therefore such a requirement is not possible. $\square$

2.3.1 Type 1 Breeding

Lemma 2.9 enables us to propose the following theorem.

Theorem 2.2. *There are no bicubic bridge graphs.*

Proof. Recall that a bridge graph can only be constructed through the application of type 1 breeding, or by performing type 1 or type 3 parthenogenesis. In the latter two cases it is clear that the resultant graph will not be bipartite, as both the parthenogenic diamond and parthenogenic triangle contain odd cycles. So we can focus solely on bridge graphs containing irreducible 1-crackers. Then, assume that there exists a bicubic bridge graph $\Gamma_D = \langle V_D, E_D \rangle$, with an irreducible 1-cracker comprising an edge $(v1, v2)$. Then v_1 must be adjacent to two other vertices, say a and b, and likewise v_2 must be adjacent to two other vertices, say c and d. Such a graph has two parents $\Gamma_1 = \langle V_1, E_1 \rangle$ and $\Gamma_2 = \langle V_2, E_2 \rangle$ that we can obtain by use of the type 1 inverse breeding operation $\mathscr{B}^{-1}(\Gamma_D, (v_1, v_2))$. In the first parent it is clear that there exists an edge (a, b), and likewise in the second parent there exists an edge (c, d) (Fig. 2.10).

Since Γ_D is bipartite, we can assign a weight of $+1$ or -1 to each vertex $v \in V_D$ in such a way that the weight of every edge is zero. Because they are both adjacent to vertex v_1, it is clear that the weight assigned to vertices a and b must be the same. Likewise, the weight assigned to vertices c and d must also be the same (but opposite to that assigned to a and b). Every edge $e \in E_D$ other than $(a, v_1), (b, v_1),$

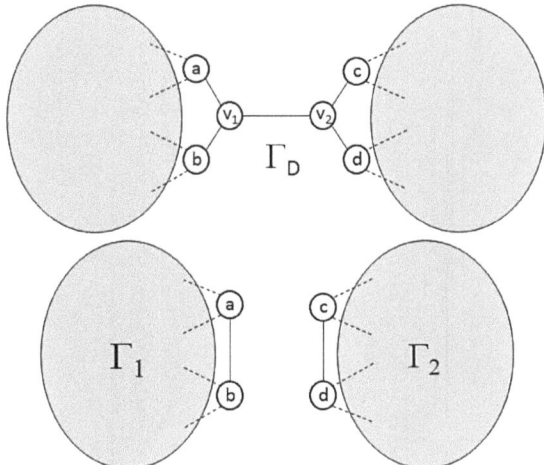

Fig. 2.10 A bicubic bridge graph, and its two parents respectively

(v_1, v_2), (c, v_2), (d, v_2) exists in either E_1 or E_2. The only edges that appear in E_1 and E_2 which are not in E_D are the edges (a, b) and (c, d). Therefore, it is possible for us to assign weights to the vertices in V_1 and V_2 in such a way that every edge in E_1 and E_2 has zero weight, other than edges $(a, b) \in E_1$ and $(c, d) \in E_2$. However, by Lemma 2.9, this is impossible, and therefore the initial assumption that there exists a bicubic bridge graph must be false. $\square$

2.3.2 Type 2 Breeding

In the proof of Theorem 2.2, it became clear that the majority of structure in a bipartite descendant must be inherited from its parents. The same theme is followed in the following three propositions, which will be used in the proof of the main theorem of this section.

Proposition 2.1. *Consider two cubic graphs* $\Gamma_1 = \langle V_1, E_1 \rangle$ *and* $\Gamma_2 = \langle V_2, E_2 \rangle$, *and a descendant* $\Gamma_D = \langle V_D, E_D \rangle$ *obtained by performing the type 2 breeding operation* $\mathscr{B}_2(\Gamma_1, \Gamma_2, a, b, c, d)$, *where* $(a, b) \in E_1$ *and* $(c, d) \in E_2$. *Then, (Figs. 2.11)*

(1) if Γ_1 *and* Γ_2 *are bipartite,* Γ_D *is also bipartite.*
(2) if at least one of Γ_1 *and* Γ_2 *is non-bipartite,* Γ_D *is non-bipartite.*

Proof. Consider first the case where Γ_1 and Γ_2 are bipartite. Then, it is possible to assign weights to the vertices in V_1 and V_2 in such a way that all edge weights in Γ_1 and Γ_2 are zero. It is clear that a must be assigned the opposite weight to

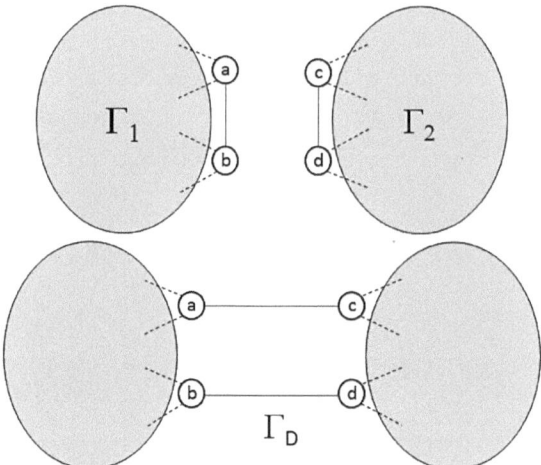

Fig. 2.11 Γ_1, Γ_2 and Γ_D as described in Proposition 2.1

b, and likewise c must be assigned the opposite weight to d. It is then possible to additionally demand that the weight assigned to a be the opposite to the weight assigned to c.

Note that $V_D = V_1 \cup V_2$. By assigning the same weights to vertices in V_D as their corresponding vertices in V_1 and V_2, all edge weights in Γ_D are zero, and therefore Γ_D is bipartite. This completes the proof of (1).

Next, consider the case where at least one of Γ_1 and Γ_2 are non-bipartite. Without loss of generality, assume that Γ_1 is non-bipartite. Then, assume that Γ_D is bipartite. Every edge $e \in E_1$ also exists in E_D except (a,b). Now, it is possible to assign weights to every vertex in V_D such that every edge in E_D has weight zero. Then, every vertex $v \in V_1$ can be assigned the same weight as the corresponding vertex in V_D, and subsequently all edges in E_1 have weight zero except for edge (a,b). From Lemma 2.9, this is impossible, and therefore the assumption that Γ_D is bipartite must be false. This completes the proof of (2). $\square$

2.3.3 Type 2 Parthenogenesis

Proposition 2.2. *Consider a cubic graph* $\Gamma = \langle V, E \rangle$ *containing a 2-cracker* $C_2 = \{(a,b),(c,d)\}$*, and a descendant* $\Gamma_D = \langle V_D, E_D \rangle$ *obtained by performing the type 2 parthenogenic operation* $\mathscr{P}_2(\Gamma,(a,b),(c,d))$*. Then, (Fig. 2.12)*

(1) if Γ *is bipartite,* Γ_D *is also bipartite.*
(2) if Γ *is non-bipartite,* Γ_D *is also non-bipartite.*

Proof. Consider first the case where Γ is bipartite. Then it is possible to assign weights to all vertices $v \in V$ such that every edge in E has weight zero. It is clear

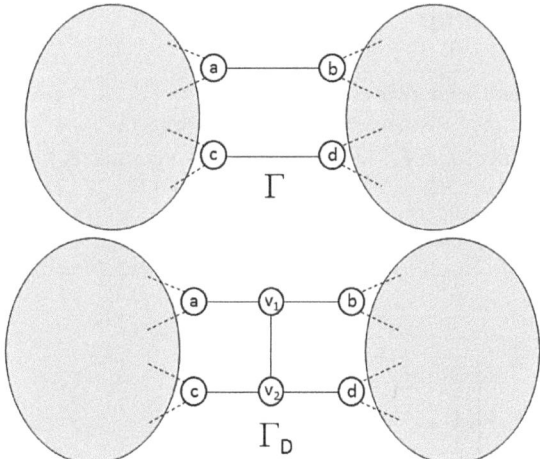

Fig. 2.12 Γ and Γ_D as described in Proposition 2.2

that a must be assigned the opposite weight to b, and likewise c must be assigned the opposite weight to d. Since C_2 is a 2-cracker, graph Γ must have been obtained from type 2 breeding. Then, from Proposition 2.1, it is clear that its two parents are also bipartite. Therefore, it is possible to request that the weight assigned to a is the opposite weight as that assigned to c, and that the weight assigned to b is the opposite weight as that assigned to d.

Since C_2 is a 2-cracker, the vertices in Γ can be divided into two subsets, V_1 and V_2, such that the only edges incident on vertices from both subsets are (a,b) and (c,d). Likewise, we can divide the vertices in Γ_D into three sets, $V_{D1} = V_1$, $V_{D2} = V_2$ and $V_{D3} = \{v_1, v_2\}$. Then, we assign weights to the vertices $v_d \in V_{D1}$ in Γ_D, in such a way that they are assigned the same weights as vertices $v \in V_1$ in Γ. Similarly, we can assign weights to the vertices $v_d \in V_{D2}$ in Γ_D, in such a way that they are assigned the *opposite* weights as vertices $v \in V_2$ in Γ. Therefore, vertices $a, b \in V_D$ will both be assigned the same weight, and vertices $c, d \in V_D$ will also be assigned the same weight, but the opposite of that assigned to a and b. Then, v_1 is assigned the same weight as c and d, and v_2 is assigned the same weight as a and b. It is clear that this assignment of weights induces all edge weights of zero in Γ_D, and therefore Γ_D is bipartite. This completes the proof of (1).

Next, consider the case where Γ is non-bipartite, and assume that Γ_D is bipartite. Then, we can assign weights to the vertices in V_D in order to obtain a zero weight for every edge in E_D. Recall the vertex divisions $V_1, V_2 \subset V$ and $V_{D1}, V_{D2}, V_{D3} \subset V_D$ used in the proof of (1). We can then assign weights to the vertices $v \in V_1$ in Γ, in such a way that they are assigned the same weight as vertices $v_d \in V_{D1}$ were. Similarly, we can assign weight to vertices $v \in V_2$ in Γ, in such a way that they are assigned the *opposite* weight as vertices $v_d \in V_{D2}$ were. It is clear that this assignment of weights induces a zero weight for every edge in E. This is a contradiction, as Γ is non-bipartite, and therefore the assumption that Γ_D is bipartite must be false. This completes the proof of (2). $\square$

2.3.4 Type 3 Breeding

Proposition 2.3. *Consider two cubic graphs* $\Gamma_1 = \langle V_1, E_1 \rangle$ *and* $\Gamma_2 = \langle V_2, E_2 \rangle$, *and a descendant* $\Gamma_D = \langle V_D, E_D \rangle$ *obtained by performing the type 3 breeding operation* $\mathscr{B}_3(\Gamma_1, \Gamma_2, v_1, v_2, a, b, c, d, e, f)$, *where* $v_1, a, b, c \in V_1$, $v_2, d, e, f \in V_2$, (a, v_1), (b, v_1), $(c, v_1) \in E_1$, *and* (d, v_2), (e, v_2), $(f, v_2) \in E_2$. *Then, (Fig. 2.13)*

(1) if Γ_1 *and* Γ_2 *are bipartite,* Γ_D *is also bipartite.*
(2) if at least one of Γ_1 *and* Γ_2 *is non-bipartite,* Γ_D *is not bipartite.*

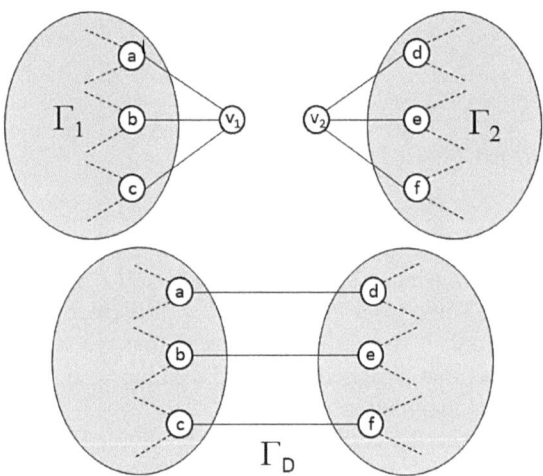

Fig. 2.13 Γ_1, Γ_2 and Γ_D as described in Proposition 2.3

Proof. Consider first the case where both Γ_1 and Γ_2 are bipartite. Then, it is possible to assign weights to the vertices in V_1 and V_2 in such a way that the all edge weights in Γ_1 and Γ_2 are zero. It is clear that the weights assigned to vertices $a, b, c \in V_1$ must all be the same, and likewise that the weights assigned to vertices $d, e, f \in V_2$ must all be the same. Without loss of generality, assume that the weights assigned to vertices a, b and c are the opposite as those assigned to vertices d, e and f. Then, in Γ_D we can assign weights to every vertex such that the weights are the same as that applied in the corresponding vertices in V_1 and V_2. The only edges in E_D that were not in $E_1 \cup E_2$ are (a, d), (b, e) and (c, f), and therefore all other edges $e \in E_D$ have zero weight. Since vertices a, b and c are assigned opposite weight than vertices d, e and f, it is clear that these new edges also have zero weight. Therefore, Γ_D is bipartite. This completes the proof of (1).

Next, consider the case where at least one of Γ_1 and Γ_2 is non-bipartite. Without loss of generality, assume that Γ_1 is non-bipartite. Then, assume that Γ_D is bipartite. Therefore, we can assign weights to the vertices in Γ_D such that all edge weights are zero. The only vertices in $V_1 \cup V_2$ that are not in V_D are v_1 and v_2. For all other vertices in $V_1 \cup V_2$, we can assign weights such that the weights are the same as the corresponding vertices in V_D. Then all edges in E_1 and E_2 must have zero weight, except those adjacent to vertices v_1 and v_2, respectively.

Now, consider the possible weight assignments for vertices a, b and c in Γ_1, as inherited from Γ_D. There are only two possibilities – either all three are assigned the same weight, or one of the three is assigned the opposite weight to the other two. If all three are assigned the same weight, then we assign v_1 the opposite weight, and consequently Γ_1 has zero weight for all edges. This is a contradiction, and so vertices a, b and c must not all be the same weight. Without loss of generality to Γ_1 being non-bipartite, assume that vertex a has opposite weight to vertices b and c. Then, if we assign v_1 the same weight as vertex a, all edges in E_1 have zero weight except edge (a, v_1). However, by Lemma 2.9, this is impossible, and therefore the assumption that Γ_D is bipartite must be false. This completes the proof of (2). □

2.3.5 Main Theorem and Discussion

From Theorem 2.2 we see that there are no bicubic bridge graphs. Then, Propositions 2.1–2.3 enable us to propose the main theorem of this section, which considers all non-bridge bicubic graphs.

Theorem 2.3. *A non-bridge descendant Γ is bicubic if and only if all of its ancestor genes are bicubic.*

Proof. Since Γ is non-bridge, no type 1 breeding operations may be used in its construction. Also, since there are no 1-crackers, no type 1 or type 3 parthenogenic operations could have been performed. Therefore, Γ can only be constructed from its ancestor genes through the combined use of type 2 breeding, type 3 breeding, and type 2 parthenogenesis. Consider the ancestor genes of Γ. If any of them are non-bipartite, then from Propositions 2.1–2.3 their descendants are also non-bipartite, and inductively we see that Γ is also non-bipartite. If all ancestor genes are bipartite, then from Propositions 2.1–2.3 their descendants are all bipartite, and inductively we see that Γ is also bipartite. □

We can now see that every bicubic graph is either a gene, or is descended from bipartite genes. This enables us to reduce the study of bicubic graphs to merely the study of bipartite genes. The following discussion considers such a problem, by searching for smaller counterexamples to Tutte's conjecture than the current known best.

Discussion 2.1 *Tutte's conjecture was that every 3-connected bicubic graph is Hamiltonian. Horton provided the first counterexample [3], and George's graph (see Fig. 2.9) is the smallest currently known counterexample [9]. In fact, George's graph is a mutant. The second-smallest known counterexample, the second Ellingham-Horton graph on 54 vertices, is also a mutant.*

From Lemma 2.8 and Theorem 2.3 we know that breeding any bipartite mutant with any other bipartite genes will output a non-Hamiltonian bipartite graph. If we restrict ourselves to using only type 3 breeding operations, then we obtain 3-connected non-Hamiltonian bipartite graphs, which constitute a counterexample to Tutte's conjecture.[2] It was shown in Remark 1.3 that there are infinitely many genes, and the Möbius ladder graphs were given as an example of such an infinite family of genes. It is easy to see by their construction that Möbius ladder graphs of size $4m + 2$ for $m = 1, 2, \ldots$ are bipartite, and therefore there are infinitely many counterexamples to Tutte's conjecture. In fact, breeding George's graph $\Gamma_G = \langle V_G, E_G \rangle$ with the 6-vertex bicubic gene $\Gamma_6^* = \langle V_6^*, E_6^* \rangle$ with the operation $\mathcal{B}_3(\Gamma_G, \Gamma_6^*, v_1, v_2, a, b, c, d, e, f)$, for any selection of $v_1, a, b, c \in V_G$, $v_2, d, e, f \in V_6^*$, $(a, v_1), (b, v_1), (c, v_1) \in E_G$ and $(d, v_2), (e, v_2), (f, v_2) \in E_6^*$ provides a 54-vertex counterexample. However, all mutant descendant counterexamples contain more vertices than their mutant ancestor gene. Therefore, it is clear that the smallest possible counterexample to Tutte's conjecture must either be a mutant, or must have all Hamiltonian genes. The following proposition eliminates the latter possibility, requiring a manual search of all descendants up to 26 vertices, rather than 48 vertices.

Proposition 2.4. *The smallest possible counterexample to Tutte's conjecture is a mutant.*

Proof. As demonstrated in Sect. 2.2.1.3, the simplest way to generate non-Hamiltonian descendants from Hamiltonian ancestor genes using type 3 breeding is to choose two genes, one containing an NH-edge, and one containing a forced edge. However, an exhaustive search of all bipartite genes up to 26 vertices found that there are no examples of either such gene for this restrictive size. Since it is impossible for type 3 breeding to introduce a new NH-edge or forced edge if the two parents do not themselves contain any, it is therefore impossible to find two such bipartite graphs whose descendant will be smaller than 50 vertices after a type 3 breeding operation is performed.   □

Such an exhaustive search to first identify all the bipartite genes up to 26 vertices, and then to check if they contain an NH-edge or a forced edge, is tractable on a modern computer in relatively short time. The same search on graphs up to 48 vertices is not. The number of connected cubic graphs containing up to 26 vertices is 2,220,297,317 [14]. The number of connected cubic graphs containing up to 48 vertices is not currently listed in literature, but Robinson and Wormald [14] gives the number containing up to 40 vertices as 9,155,199,875,057,748,089, or roughly 4×10^9 times as many graphs as those containing up to 26 vertices. Judging by the exponential growth in the number of graphs for larger vertices, by the time we consider 48-vertex cubic graphs it is likely that this number would increase to approximately 5×10^{15} times as many graphs. That we can avoid considering these graphs is an example of the advantage gained from the present theory.

---

[2] In fact, any breeding operations other than type 3 breeding prevents 3-connectedness in all subsequence descendants, since cubic crackers cannot be destroyed by further breeding. Therefore, if a 3-connected descendant is desired, only type 3 breeding operations may be used.

Discussion 2.2 *The famous Horton graph [3] is not a mutant, but rather a non-Hamiltonian non-bridge descendant containing four ancestor genes, three of which are a 32-vertex bipartite gene containing a forced edge, and fourth being the 6-vertex gene Γ_6^*. The first Ellingham-Horton graph is also not a mutant, but rather a descendant containing three ancestor genes, two of which are the same 32-vertex bipartite gene containing a forced edge, and an 18-vertex bipartite gene containing an NH-edge-pair (i.e. an NH-edge-set of cardinality 2). In fact, this 18-vertex gene is the smallest such bipartite gene to contain an NH-edge-pair.*

Due to computational restriction, we do not currently know whether any bipartite genes, of sizes 28 or 30, contain an NH-edge or a forced edge. If not, the 32-vertex gene used in the construction of the Horton graph and the first Ellingham-Horton graph are the smallest example of such a bipartite gene.

The Horton graph, the first Ellingham-Horton graph, and their ancestors are displayed in Figs. 2.14–2.15.

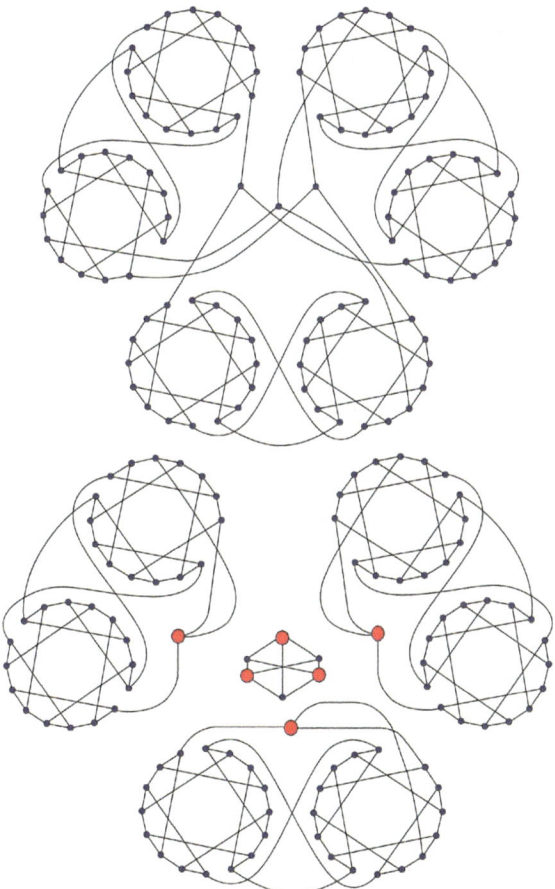

Fig. 2.14 The Horton graph, and its ancestor genes

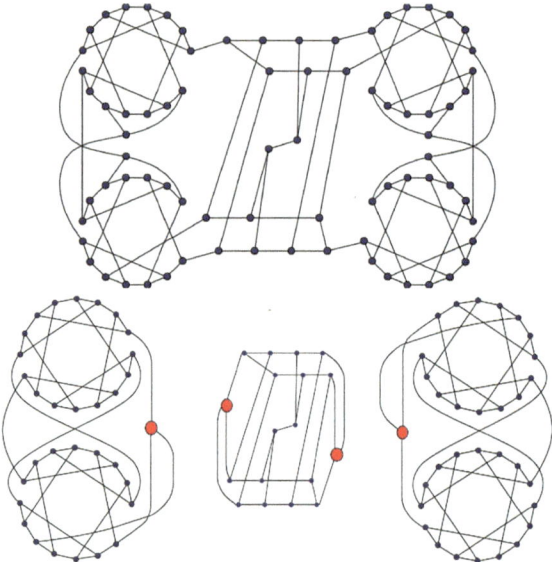

Fig. 2.15 The first Ellingham-Horton graph, and its ancestor genes

2.4 Planarity

A *planar* graph is one that can be visually represented in such a way that all edges intersect only at vertex points. That is, none of the edges cross any other edges. The inheritance of planarity in graphs from their minors has been studied [12] and many other criteria have been given (e.g. see Robertson and Seymour [13]). Planar graphs were the subject of Tait's conjecture that all 3-connected cubic planar graphs are Hamiltonian, which was first disproved by Tutte [16]. Holton and McKay proved that the smallest possible counterexamples contain 38 vertices [11]. There are six such graphs up to isomorphism, known as the Barnette-Bośak-Lederberg graphs. Unlike Tutte's conjecture, where the smallest counterexamples are mutants, the Barnette-Bośak-Lederberg graphs are all descendants. They will be discussed in some detail at the end of this section.

In this section, we demonstrate that planarity is also inherited by descendants through the six breeding operations, and retained by the six inverse operations. Certainly planar genes exists, of which Γ_4^* is the smallest. Planar mutants also exist (and constitute counterexamples to Tait's conjecture), of which the smallest given in literature is the Faulkner-Younger graph on 42 vertices [5].[3] These two graphs are displayed in Fig. 2.16.

The upcoming six propositions will be aided by three simple properties of planar graphs that can be easily verified.

---

[3] Note that the Grinberg graph on 42 vertices [10] is an equally small example of a planar mutant.

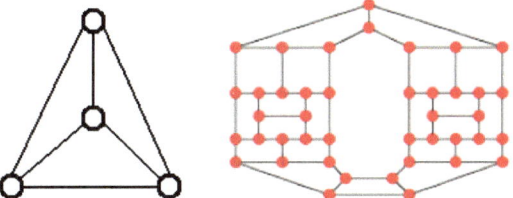

Fig. 2.16 The two smallest planar Hamiltonian and non-Hamiltonian genes, Γ_4^* and the Faulkner-Younger graph 42 respectively

Pr.1 In a planar graph, if any edge is removed, the resulting graph is planar. Likewise, if any vertex is removed (and the adjacent edges are therefore removed as well), the resulting graph is planar.

Pr.2 In a planar graph, a vertex may be placed inside any edge (such that if an edge (a,b) has vertex c placed inside it, the edge (a,b) is replaced with two edges (a,c) and (c,b)), and the resulting graph is planar. Equivalently, any degree 2 vertex can be removed in the sense that a single edge is placed between its adjacent vertices, and the resulting graph is planar.

Pr.3 In a planar graph, any vertex may be replaced with a planar subgraph, in such a way that each edge that was adjacent to the replaced vertex becomes adjacent to an external vertex in the planar subgraph, and in such a way that these edges do not overlap. Then, the resulting graph is planar. Equivalently, any connected subgraph in a planar graph can be replaced with a vertex, such that all external edges to the subgraph become adjacent to the new vertex, and the resulting graph is planar.

2.4.1 Type 1 Breeding

Proposition 2.5. *Consider two graphs $\Gamma_1 = \langle V_1, E_1 \rangle$ and $\Gamma_2 = \langle V_2, E_2 \rangle$. Suppose there is an edge $(a,b) \in E_1$ and an edge $(c,d) \in E_2$. Then, consider the descendant graph $\Gamma_D = \langle V_D, E_D \rangle$ (where $V_D = V_1 \cup V_2 \cup \{v_1, v_2\}$) obtained from the type 1 breeding operation $\mathscr{B}_1(\Gamma_1, \Gamma_2, (a,b), (c,d))$. These graphs are displayed in Fig. 2.17.*

(1) If Γ_1 and Γ_2 are both planar, then Γ_D is planar.
(2) If Γ_D is planar, then Γ_1 and Γ_2 are both planar.

Proof. First, consider the case where Γ_1 and Γ_2 are both planar. Then, by Pr.2 we can add a vertex v_1 inside edge (a,b), and another vertex v_2 inside edge (c,d), and the two resulting graphs are planar. Call these two new graphs $\overline{\Gamma}_1$ and $\overline{\Gamma}_2$. Then, consider the introduction of a new vertex v_3 and edge (v_1, v_3) introduced to $\overline{\Gamma}_1$ to obtain $\widetilde{\Gamma}_1$. This is equivalent to replacing vertex v_1 with the planar subgraph $\Gamma_S = \langle \{v_1, v_3\}, \{(v_1, v_3)\} \rangle$ (where edges (a, v_1) and (b, v_1) remain unchanged), and

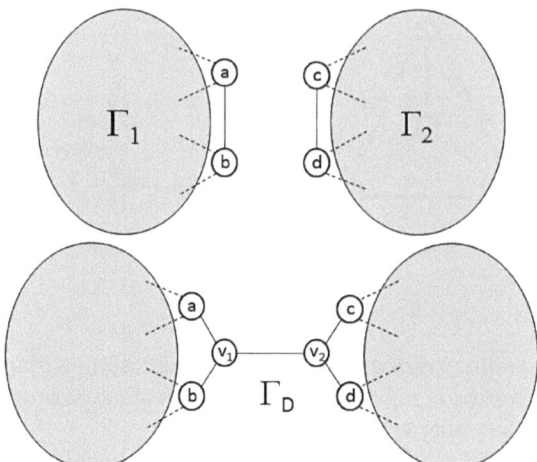

Fig. 2.17 Γ_1, Γ_2 and Γ_D as described in Proposition 2.5

so by Pr.3, $\widetilde{\Gamma_1}$ is planar. Graphs $\widetilde{\Gamma_1}$ and $\overline{\Gamma}_2$ are displayed in Fig. 2.18. Then, we can replace v_3, in $\widetilde{\Gamma_1}$, with the graph $\overline{\Gamma}_2$, where edge (v_1, v_3) is replaced by the new edge (v_1, v_2). The resultant graph is equivalent to Γ_D, and so by Pr.3, Γ_D is planar. This completes the proof of (1).

Next, consider the case where Γ_D is planar. Then by Pr.1 we can delete the 1-cracker (v_1, v_2) to separate the graph into two planar subgraphs, $\overline{\Gamma}_1$ and $\overline{\Gamma}_2$. Then, we can delete both vertices v_1 and v_2 in the sense that edges (a, v_1) and (b, v_1) are replaced with edge (a, b) in Γ_1, and edges (c, v_2) and (d, v_2) are replaced with edge (c, d) in Γ_2. The two resulting subgraphs are equivalent to Γ_1 and Γ_2, and so by Pr.2, they are both planar. This completes the proof of (2). □

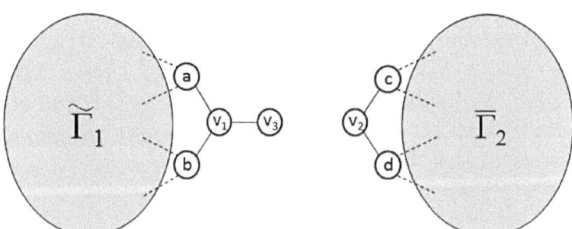

Fig. 2.18 $\widetilde{\Gamma}_1$ and $\overline{\Gamma}_2$ as described in the proof of Proposition 2.5

2.4.2 Type 2 Breeding

Proposition 2.6. *Consider two graphs* $\Gamma_1 = \langle V_1, E_1 \rangle$ *and* $\Gamma_2 = \langle V_2, E_2 \rangle$. *Suppose there is an edge* $(a, b) \in E_1$ *and an edge* $(c, d) \in E_2$. *Then, consider the descendant graph* $\Gamma_D = \langle V_D, E_D \rangle$ *(where* $V_D = V_1 \cup V_2$*) obtained from the type 2 breeding operation* $\mathscr{B}_2(\Gamma_1, \Gamma_2, a, b, c, d)$. *These graphs are displayed in Fig. 2.19.*

(1) If Γ_1 *and* Γ_2 *are both planar, then* Γ_D *is planar.*
(2) If Γ_D *is planar, then* Γ_1 *and* Γ_2 *are both planar.*

Proof. First, consider the case where Γ_1 and Γ_2 are both planar. Then, in Γ_1, by Pr.2, we can add a vertex, say v_1, inside edge (a, b) in such a way that the edge is replaced by edges (a, v_1) and (b, v_1), to obtain a planar graph $\overline{\Gamma}_1$. In Γ_2, by Pr.1, we can delete edge (c, d) to obtain a planar graph $\widetilde{\Gamma}_2$. Graphs $\overline{\Gamma}_1$ and $\widetilde{\Gamma}_2$ are displayed in Fig. 2.20. Finally, we can replace vertex v_1 with the graph $\widetilde{\Gamma}_2$, where edges (a, v_1) and (b, v_1) are replaced with new edges (a, c) and (b, d) respectively. It is clear that edges (a, c) and (b, d) need not intersect (if they do, $\widetilde{\Gamma}_2$ can just be turned upside-down). The resultant graph is equivalent to Γ_D, and so by Pr.3, Γ_D is planar. This completes the proof of (1).

Next, consider the case where Γ_D is planar. Recall that $V_D = V_1 \cup V_2$. Consider the subgraph $\widetilde{\Gamma}_2$ containing all vertices in V_2, and all edges incident only on vertices in V_2. It is clear that $\widetilde{\Gamma}_2$ is connected. Then, by Pr.3, within Γ_D we can replace subgraph $\widetilde{\Gamma}_2$ by a vertex v_1, introducingedges (a, v_1) and (b, v_1), to obtain a planar graph $\overline{\Gamma}_1$. Then, by removing vertex v_1, replacing edges (a, v_1) and (b, v_1) with edge (a, b), we obtain Γ_1. By Pr.2, Γ_1 is planar. Using an equivalent argument it can easily be seen that Γ_2 is also planar. This completes the proof of (2). □

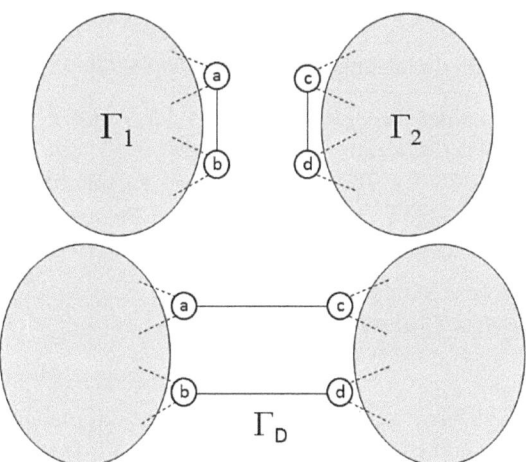

Fig. 2.19 Γ_1, Γ_2 and Γ_D as described in Proposition 2.6

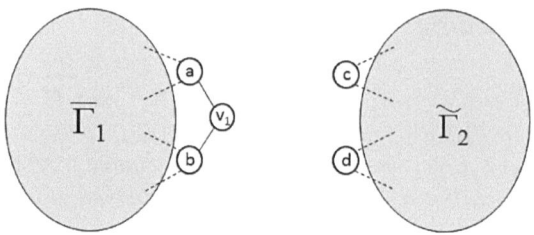

Fig. 2.20 $\overline{\Gamma}_1$ and $\widetilde{\Gamma}_2$ as described in the proof of Proposition 2.6

2.4.3 Type 3 Breeding

The following lemma is used in the proof of Proposition 2.7.

Lemma 2.10. *Any 3-cracker can be drawn with nonintersecting edges.*

Proof. Consider the case where a 3-cracker (joining two subgraphs Γ_1 and Γ_2) is drawn with straight edges. If none are intersecting, we are done. Now, consider the alternative case, where at least two of the edges intersect. There are three possibilities:

(1) All three edges intersect the other two edges.
(2) Two edges are nonintersecting with respect to each other, but both intersect the third edge.
(3) Two edges intersect each other, but neither edge intersects the third.

For (1), simply turning one of the subgraphs upside-down gives nonintersecting edges. For (2), the edge that intersects both other edges can be redirected around one subgraph to give nonintersecting edges. For (3), turning one of the subgraphs "upside-down" reduces the situation to (2). ☐

The above three situations and solutions are displayed in Fig. 2.21.

Proposition 2.7. *Consider two graphs $\Gamma_1 = \langle V_1, E_1 \rangle$ and $\Gamma_2 = \langle V_2, E_2 \rangle$. Suppose there is a vertex $v_1 \in V_1$ adjacent to vertices $a, b, c \in V_1$, and a vertex $v_2 \in V_2$ adjacent to vertices $d, e, f \in V_2$. Then, consider the descendant graph $\Gamma_D = \langle V_D, E_D \rangle$ (where $V_D = \{V_1 \setminus v_1\} \cup \{V_2 \setminus v_2\}$) obtained from the type 3 breeding operation $\mathcal{B}_3(\Gamma_1, \Gamma_2, v_1, v_2, a, b, c, d, e, f)$. These graphs are displayed in Fig. 2.22.*

(1) If Γ_1 and Γ_2 are both planar, then Γ_D is planar.
(2) If Γ_D is planar, then Γ_1 and Γ_2 are both planar.

Proof. First, consider the case where Γ_1 and Γ_2 are both planar. Then, by Pr.1, we can remove vertex v_2 and edges (d, v_2), (e, v_2) and (f, v_2) from Γ_2 to obtain a planar graph $\widetilde{\Gamma}_2$, displayed in Fig. 2.23. Then, we can replace vertex v_1 in Γ_1 with subgraph $\widetilde{\Gamma}_2$, in such a way that edges (a, v_1), (b, v_1) and (c, v_1) are replaced with new

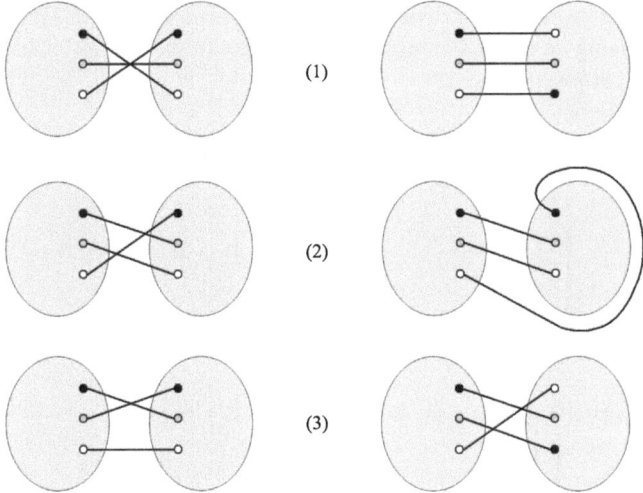

Fig. 2.21 The three possible types of intersecting 3-crackers detailed in Lemma 2.10, and how the graphs can be redrawn to avoid the intersecting edges

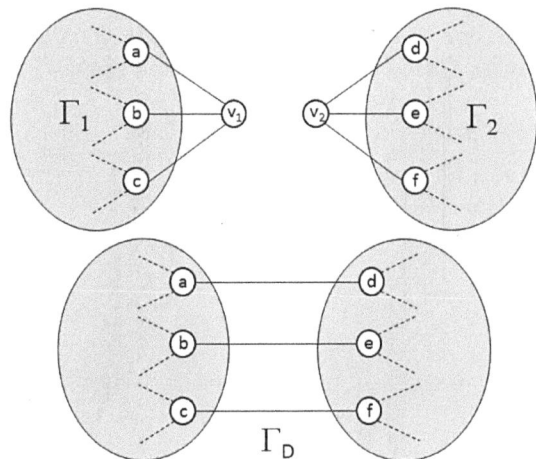

Fig. 2.22 Γ_1, Γ_2 and Γ_D as described in Proposition 2.7

edges (a,d), (b,e) and (c,f). The resultant graph is equivalent to Γ_D, and from Lemma 2.10 the edges need not overlap. Therefore, by Pr.3, Γ_D is planar. This completes the proof of (3).

Next, assume that Γ_D is planar. Recall that $V_D = \{V_1 \setminus v_1\} \cup \{V_2 \setminus v_2\}$. Consider the subgraph $\widetilde{\Gamma_2}$ containing all vertices in $\{V_2 \setminus v_2\}$, and all edges adjacent only to vertices in $\{V_2 \setminus v_2\}$. It is clear that $\widetilde{\Gamma_2}$ is connected. Then, within Γ_D, we can replace subgraph $\widetilde{\Gamma_2}$ by a vertex v_1, replacing edges (a,d), (b,e) and (c,f) by edges (a,v_1),

(b,v_1) and (c,v_1) respectively. The resultant graph is equivalent to Γ_1, and so by Pr.3, Γ_1 is planar. Using an equivalent argument it is easy to see that Γ_2 is also planar. This completes the proof of (2). $\quad\square$

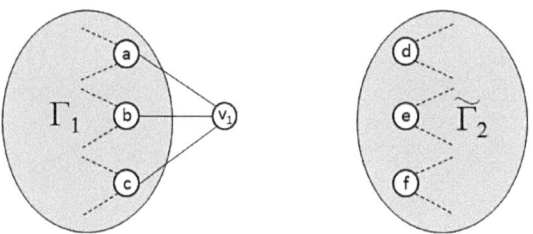

Fig. 2.23 $\widetilde{\Gamma_1}$ and $\widetilde{\Gamma_2}$ as described in the proof of Proposition 2.7

2.4.4 Type 1 Parthenogenesis

Proposition 2.8. *Consider a descendant graph* $\Gamma_1 = \langle V_1, E_1 \rangle$ *containing a 1-cracker comprising edge* (a,b). *Then, consider the descendant graph* $\Gamma_D = \langle V_D, E_D \rangle$ *(where* $V_D = V_1 \cup \{v_1, v_2, v_3, v_4\}$*) obtained from the type 1 parthenogenic operation* $\mathscr{P}_1(\Gamma_1, (a,b))$. *These graphs are displayed in Fig. 2.24.*

(1) If Γ_1 *is planar, then* Γ_D *is planar.*
(2) If Γ_D *is planar, then* Γ_1 *is planar.*

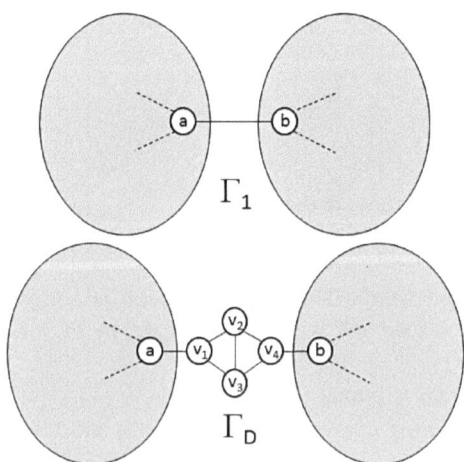

Fig. 2.24 Γ and Γ_D as described in Proposition 2.8

Proof. First, consider the case where Γ_1 is planar. Then, by Pr.2, we can introduce a vertex v inside edge (a,b), in such a way that edge (a,b) is replaced with new edges (a,v) and (b,v), to obtain $\overline{\Gamma}_1$, displayed in Fig. 2.25. Then, we can replace vertex v in $\overline{\Gamma}_1$ with the parthenogenic diamond, which is a planar subgraph, in such a way as to replace edges (a,v) and (b,v) with new edges (a,v_1) and (b,v_4) respectively. The resultant graph is equivalent to Γ_D, and therefore by Pr.3, Γ_D is planar. This completes the proof of (1).

Next, assume that Γ_D is planar. We can replace the parthenogenic diamond with a vertex v_1, introducing edges (a,v_1) and (b,v_1), to obtain $\overline{\Gamma}_1$. By Pr.3, $\overline{\Gamma}_1$ is planar. Then, by removing vertex v_1, replacing edges (a,v_1) and (b,v_1) with edge (a,b), we obtain Γ_1. By Pr.2, Γ_1 is planar. This completes the proof of (2). □

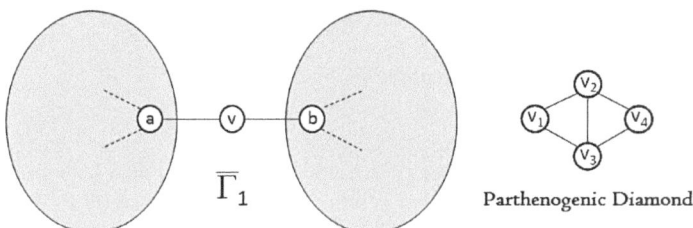

$\overline{\Gamma}_1$ Parthenogenic Diamond

Fig. 2.25 $\overline{\Gamma}_1$ and the parthenogenic diamond as described in the proof of Proposition 2.8

2.4.5 Type 2 Parthenogenesis

Proposition 2.9. *Consider a descendant graph $\Gamma_1 = \langle V_1, E_1 \rangle$ containing a 2-cracker comprising edges (a,b) and (c,d). Then, consider the descendant graph $\Gamma_D = \langle V_D, E_D \rangle$ (where $V_D = V_1 \cup \{v_1, v_2\}$) obtained from the type 2 parthenogenic operation $\mathscr{P}_2(\Gamma_1, (a,b), (c,d))$. These graphs are displayed in Fig. 2.26.*

(1) If Γ_1 is planar, then Γ_D is planar.
(2) If Γ_D is planar, then Γ_1 is planar.

Proof. First, consider the case where Γ_1 is planar. Then, we can introduce a vertex v_1 inside edge (a,b), in such a way that edge (a,b) is replaced with new edges (a,v_1) and (b,v_1), and another vertex v_2 inside edge (c,d), in such a way that edge (c,d) is replaced with new edges (c,v_2) and (d,v_2), to obtain $\overline{\Gamma}_1$, displayed in Fig. 2.27. By Pr.2, $\overline{\Gamma}_1$ is planar. Then, since (a,b) and (c,d) form a 2-cracker in Γ_1, it is clear that an edge (v_1, v_2) may be introduced without intersecting any other edges in $\overline{\Gamma}_1$. The resultant graph is equivalent to Γ_D, and therefore Γ_D is planar. This completes the proof of (1).

Next, assume that Γ_D is planar. By Pr.1, we can remove edge (v_1, v_2) to obtain a planar graph $\overline{\Gamma}_1$. Then, we can remove vertices v_1 and v_2, such that edges (a,v_1)

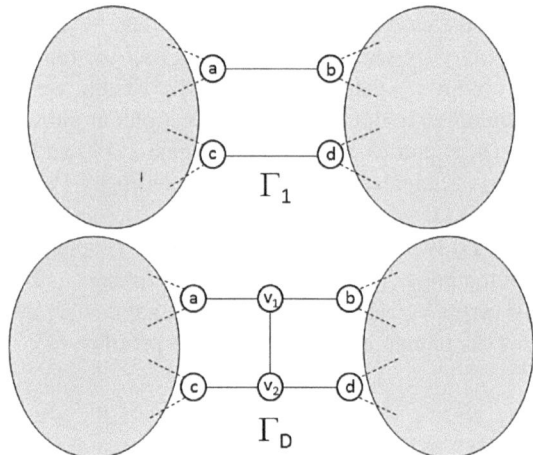

Fig. 2.26 Γ and Γ_D as described in Proposition 2.9

and (b,v_1) are replaced with edge (a,b), and edges (c,v_2) and (d,v_2) are replaced with edge (c,d). The resultant graph is equivalent to Γ_1, and so by Pr.2, Γ_1 is planar. This completes the proof of (2).   □

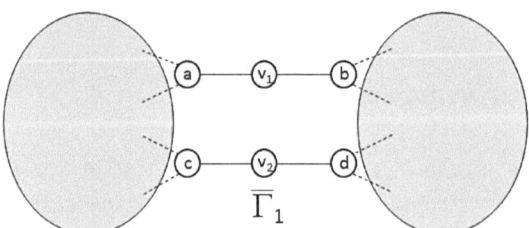

Fig. 2.27 $\overline{\Gamma}_1$ as described in the proof of Proposition 2.9

2.4.6 Type 3 Parthenogenesis

Proposition 2.10. *Consider a descendant graph* $\Gamma_1 = \langle V_1, E_1 \rangle$ *containing a 1-cracker comprising edge* (a,b). *Suppose that vertex a is also adjacent to vertices* $c, d \in V_1$. *Then, consider the descendant graph* $\Gamma_D = \langle V_D, E_D \rangle$ *(where* $V_D = V_1 \cup \{v_1, v_2\}$*) obtained from the type 3 parthenogenic operation* $\mathscr{P}_3(\Gamma_1, a)$. *These graphs are displayed in Fig. 2.28.*

(1) If Γ_1 *is planar, then* Γ_D *is planar.*
(2) If Γ_D *is planar, then* Γ_1 *is planar.*

Proof. First, consider the case where Γ_1 is planar. Then, we can replace vertex a with a parthenogenic triangle subgraph, which is planar, in such a way that edges (a,c) and (a,d) are replaced by new edges (v_1,c) and (v_2,d), and edge (a,b) remains unchanged. The resultant graph is equivalent to Γ_D, and so by Pr.3, Γ_D is planar. This completes the proof of (1).

Next, assume that Γ_D is planar. By Pr.3, we can replace the parthenogenic triangle subgraph with a vertex a, replacing edges (a,b), (v_1,c) and (v_2,d) with edges (a,b), (a,c) and (a,d). The resultant graph is equivalent to Γ_1, and so by Pr.3, Γ_1 is planar. This completes the proof of (2). $\square$

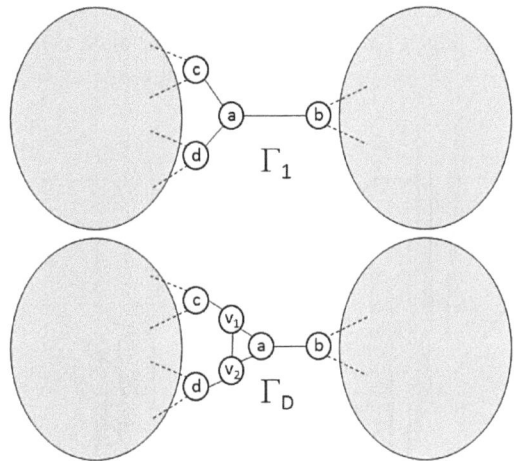

Fig. 2.28 Γ and Γ_D as described in Proposition 2.10

2.4.7 Main Theorem and Discussion

Propositions 2.5–2.10 enable us to give the main theorem of this section.

Theorem 2.4. *A descendant graph Γ_D is planar if and only if all of its ancestor genes are planar.*

Proof. From Theorem 1.2, Γ_D has a unique complete family of ancestor genes. Each of their child graphs must be constructed by one of the six breeding operations. By Propositions 2.5–2.10, these child graphs will all be planar if and only if all of the ancestor genes are planar. By induction, we can continue this argument until we arrive at Γ_D. $\square$

As discussed at the beginning of this section, the smallest counterexamples to Tait's conjecture are the Barnette-Bośak-Lederberg graphs on 38 vertices. There are

six of these graphs, and all six are descendants of the same three Hamiltonian ancestor genes. The only distinction between the six graphs is the ordering used in the
type 3 breeding operation.

Two of the ancestor genes, which we call Γ_1 and Γ_2, are in fact the same graph. As
can be seen, the Barnette-Bośak-Lederberg graph displayed in Fig. 2.29 is produced
by two type 3 breeding operations. First, the operation $\mathscr{B}_3(\Gamma_1,\Gamma_3,h,a,i,j,k,d,b,c)$
is performed to produce a descendant, say Γ_D, and then the operation $\mathscr{B}_3(\Gamma_2,\Gamma_D,$
$l,e,m,n,o,g,b,f)$ is performed.

This descendant is non-Hamiltonian because two forced edges, one in each of
Γ_1 and Γ_2, are connected with a NH-edge-pair in Γ_3. The forced edge in Γ_1 is edge
(h,k), and the forced edge in Γ_2 is (l,o). There are many NH-edge-pairs in Γ_3, but
due to isomorphism, only one needs to be considered. In Fig. 2.29 we select the NH-
edge-pair comprising edges (a,c) and (e,f). Then, once the two type 3 breeding
operations are performed, the resultant descendant contains edges (c,k) and (f,o).

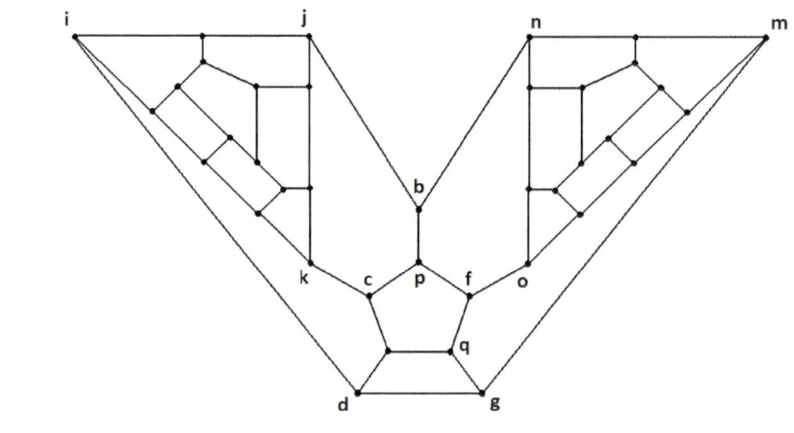

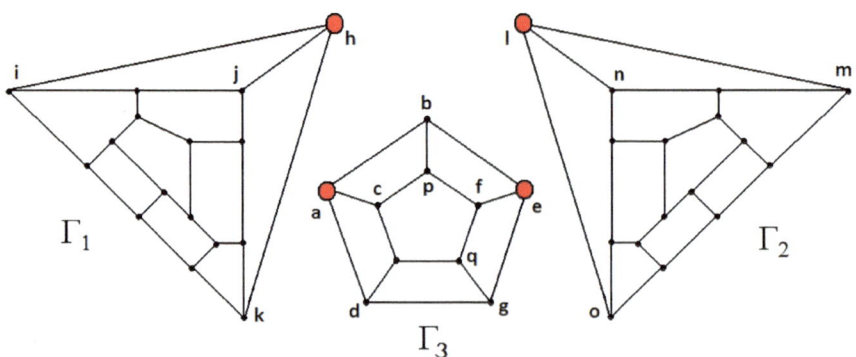

Fig. 2.29 One of the six Barnette-Bośak-Lederberg graph, and its ancestor genes

Therefore, any Hamiltonian cycle in the descendant corresponds, in part, to a Hamiltonian cycle in Γ_3 containing edges (a, c) and (e, f), which does not exist.

The other five Barnette-Bośak-Lederberg graphs can be produced by altering the ordering as follows:

- First $\mathscr{B}_3(\Gamma_1, \Gamma_3, h, a, i, j, k, b, d, c)$, then $\mathscr{B}_3(\Gamma_2, \Gamma_D, l, e, m, n, o, b, g, f)$.
- First $\mathscr{B}_3(\Gamma_1, \Gamma_3, h, a, i, j, k, d, b, c)$, then $\mathscr{B}_3(\Gamma_2, \Gamma_D, l, e, m, n, o, b, g, f)$.
- First $\mathscr{B}_3(\Gamma_1, \Gamma_3, h, a, i, j, k, b, d, c)$, then $\mathscr{B}_3(\Gamma_2, \Gamma_D, l, f, m, n, o, p, q, e)$.
- First $\mathscr{B}_3(\Gamma_1, \Gamma_3, h, a, i, j, k, d, b, c)$, then $\mathscr{B}_3(\Gamma_2, \Gamma_D, l, f, m, n, o, q, p, e)$.
- First $\mathscr{B}_3(\Gamma_1, \Gamma_3, h, a, i, j, k, d, b, c)$, then $\mathscr{B}_3(\Gamma_2, \Gamma_D, l, f, m, n, o, p, q, e)$.

Note in the latter three graphs, vertex f is removed in Γ_D rather than vertex e. In these instances, the descendant contains edges (c, k) and (e, o), which also demand the NH-edge-pair be traversed in any corresponding Hamiltonian cycle in Γ_3.

Obviously there are more than six combinations available, however, due to isomorphism, any other combination produces a graph which is isomorphic to one of the six Barnette-Bośak-Lederberg graphs.

The ancestor genes Γ_1 and Γ_2 are both the same 16-vertex gene, and contain a forced edge. This gene is the only 16-vertex planar gene to contain a forced edge, and no smaller planar genes contain a forced edge. The parent gene Γ_3 is a 10-vertex planar gene that contains an NH-edge-pair. This gene is the only 10-vertex planar gene to contain a NH-edge-pair, and no smaller genes contain an NH-edge-pair.

There are no planar genes with 22 vertices or less that contain an NH-edge, so it is not possible to obtain a descendant with fewer than 38 vertices by breeding Γ_1 with a planar gene containing an NH-edge. It is therefore clear that no 3-connected planar descendants with fewer than 38 vertices are non-Hamiltonian. Of course, the result given in [11] is stronger, as it also eliminates the possibility of planar mutants with fewer than 38 vertices.

2.5 Summary and Conclusions

We can summarise the results of this chapter with the following corollary.

Corollary 2.4. *For a descendant to be:*

- **Hamiltonian***: It is necessary for all ancestor genes to be Hamiltonian.*
- **Bipartite***: It is necessary and sufficient for all ancestor genes to be bipartite, unless the descendant is a bridge graph, in which case it is never bipartite.*
- **Planar***: It is necessary and sufficient for all ancestor genes to be planar.*

Proof. Follows immediately from Lemma 2.8 and Theorems 2.3 and 2.4. $\square$

To determine the bipartiteness or planarity of a descendant, one could, in theory, decompose the descendant into its ancestor genes, and determine the bipartiteness or planarity of these genes instead. Although this would likely represent a saving in time if the ancestor genes were known in advance, the process of identifying the ancestor genes is an $O(N^4)$ process in general, and therefore such a decomposition is useful only for problems where the best known algorithms are slow, or instances in which the ancestor genes are easy to identify. Of course, all of the results in this chapter are valid for any complete family of ancestors, not only a complete family of ancestor genes. In some cases, identifying a complete family of ancestors may be much more efficient than decomposing entirely into genes, and this may still represent a large saving in time for identifying properties of the original descendant graph.

The results of Sect. 2.2 give rise to a natural decomposition-based heuristic for the Hamiltonian cycle problem. Bridge graphs can be easily detected, so only non-bridge graphs need to be considered. For a given descendant, we can identify its ancestor genes, and then solve the Hamiltonian cycle problem for each of them, using whatever state of the art algorithms are available for solving cubic graphs (e.g. see Eppstein [4] or Baniasadi et al. [1]). Although the total solving time would still be exponential, the exponent would be proportional to the size of the largest ancestor gene, which in some cases might be much smaller than the descendant. For large mutant descendants, this represents a very significant saving in time. However, non-bridge NMNH cubic graphs would remain undetected by this heuristic. How best to address these graphs, and how many of them may be detected by more direct methods, is a topic for future research. It is worth noting that non-bridge non-Hamiltonian cubic graphs (whether mutant descendants or otherwise) are conjectured to be very rare compared to the set of non-Hamiltonian cubic graphs [6], which in turn are known to be very rare compared to the set of all cubic graphs [15].

If the problem of finding *every* Hamiltonian cycle in a cubic graph is considered, a natural decomposition algorithm emerges. First, identify all the ancestor genes, then find every Hamiltonian cycle in each of the ancestor genes. If any 2-crackers exist in the descendant, cycles that do not traverse the corresponding edges in the ancestor genes can be discarded. Finally, consider all combinations of Hamiltonian cycles in the ancestor genes to see which combinations "match up". That is, which combinations induce the same edge crossings in the descendant's cubic crackers. Discard the combinations that do not match up, and those that remain constitute all of the Hamiltonian cycles in the descendant.

To date, Hamiltonicity, bipartiteness and planarity are the only graph theoretic properties that have been investigated in the context of descendants and their ancestor genes. The investigation of other such properties is a subject for future research.

References

1. Baniasadi, P., Ejov, V., Filar, J.A., Haythorpe, M., Rossomakhine, S.: Deterministic "Snakes and Ladders" Heuristic for the Hamiltonian Cycle Problem. Mathematical Programming Computation **6**(1): 55–75, (2014)
2. Barnette, D.: In: Tutte, W.T. (ed.) Recent Progress in Combinatorics: Proceedings of the Third Waterloo Conference on Combinatorics, May 1968. Conjecture 5 Academic, New York (1969)

3. Bondy, J.A., Murty, U.S.R.: Graph Theory with Applications, pp. 61 and 242. North Holland, New York (1976)
4. Eppstein, D.: The traveling salesman problem for cubic graphs. In: Dehne, F., Sack, J.-R., Smid, M. (eds.) Algorithms and Data Structures. Lecture Notes in Computer Science, vol. 2748, pp. 307–318. Springer, Berlin/Heidelberg (2003)
5. Faulkner, G.B., Younger, D.H.: Non-Hamiltonian cubic planar maps. Discret. Math. **7**, 67–74 (1974)
6. Filar, J.A., Haythorpe, M., Nguyen, G.T.: A conjecture on the prevalence of cubic bridge graphs. Discuss. Math. Graph Theory **30**(1), 175–179 (2010)
7. Garey, M.R., Johnson, D.S.: Computers and Intractibility: A Guide to the Theory of NP-Completeness. W.H. Freeman, New York (1979)
8. Garey, M.R., Johnson, D.S., Tarjan, R.E.: The planar Hamiltonian circuit problem is NP-complete. SIAM J. Comput. **5**, 704–714 (1976)
9. Georges, J.P.: Non-Hamiltonian bicubic graphs. J. Comb. Theory Ser. B **46**, 121–124 (1989)
10. Grinberg, E.J.: Plane homogeneous graphs of degree three without Hamiltonian circuits. Latv. Math. Yearbook Izdat. Zinat. Riga **4**, 51–58 (1968)
11. Holton, D.A., McKay, B.D.: The smallest non-Hamiltonian 3-connected cubic planar graphs have 38 vertices. J. Comb. Theory Ser. B **45**(3), 305–319 (1988)
12. Kuratowski, C.: Sur le problème des courbes gauches en topologie. Fundam. Math. **15**, 217–283 (1930)
13. Robertson, N., Seymour, P.: Graph minors. XX. Wagner's conjecture. J. Comb. Theory Ser. B **92**(2), 325–357 (2004)
14. Robinson, R., Wormald, N.: Number of cubic graphs. J. Graph Theory **7**, 463–467 (1983)
15. Robinson, R., Wormald, N.: Almost all regular graphs are Hamiltonian. Random Struct. Algorithms **5**(2), 363–374 (1994)
16. Tutte, W.T.: On Hamiltonian circuits. J. Lond. Math. Soc. (2nd ser.) **21**(2), 98–101 (1946)

Chapter 3
Uniqueness of Ancestor Genes

Abstract In Chap. 1 we stated the result that every graph has a unique complete family of ancestor genes. The result is proved in detail in this chapter. The proof is lengthy and is therefore broken up into several intermediate steps. We first show that it is sufficient to prove the result for graphs with no parthenogenic objects. We then consider all possible methods of decomposing a graph into three components via the inverse operations and show that the components so obtained are always the same regardless of the inverse operations used. Next we prove that any complete family of ancestor genes for a graph has cardinality which is a fixed constant for that graph. We then proceed to prove that for any descendant without parthenogenic objects, it is possible to isolate at least two genes with single inverse breeding operations. Finally, we use each of these results to prove the uniqueness theorem.

3.1 Introduction

At the conclusion of Chap. 1, we announced the following, main, result.

Theorem 3.1. *Any descendant* Γ_D *has a unique family of ancestor genes.*

The interpretation of Theorem 3.1 is that any descendant graph may only be constructed from a particular family of ancestor genes through the use of the six breeding operations. However, note that all of the breeding operations are invertible (see Definitions 1.9 and 1.10), and from Theorem 1.1 we know that any sequence of valid inverse operations of sufficient length leads to a family of ancestor genes. Therefore, Theorem 3.1 is equivalent to stating that for a given descendant, any valid decomposition into a family of ancestor genes via inverse operations will result in the same family of ancestor genes as for any other valid decomposition. This chapter is dedicated to the proof of Theorem 3.1, with some of the more tedious details left for Appendix A.

© Springer International Publishing Switzerland 2016

P. Baniasadi et al., *Genetic Theory for Cubic Graphs*, SpringerBriefs
in Operations Research, DOI 10.1007/978-3-319-19680-0_3

In addition to the definitions and results taken from Chap. 1, we also introduce the following new definitions which 'will be used extensively throughout this chapter.

Definition 3.1. If an edge e is contained in at least one cubic cracker of a graph Γ, then we say that e is a *cubic cracker edge*.

Definition 3.2. If the deletion of a set of edges from a (possibly disconnected) graph results in more connected components than were originally present, we say that the edges *separate* those (newly resulting) connected components.

Definition 3.3. Any graph obtained after a *sequence of inverse operations* performed on a cubic graph Γ is an *ancestor* of Γ.

A sequence of inverse operations on Γ involves an inverse operation on Γ, then on ancestors of Γ, next on ancestors of ancestors of Γ and so on. When such a sequence of inverse operations is performed, the end result is a family of ancestors which are, in turn, connected cubic graphs. Rather than thinking of them as individual graphs, it will be convenient to view a family of ancestors as a single (disconnected) graph made up of the union of the ancestors. We now define such a graph.

Definition 3.4. Suppose a sequence of inverse breeding and inverse parthenogenic operations on Γ results in a (disconnected) graph $\Gamma^{(\cdot)}$. Each connected component of $\Gamma^{(\cdot)}$ is an *ancestor* of Γ, and we refer to $\Gamma^{(\cdot)}$ as an *ancestral graph* of Γ. We say that the family of all connected components of $\Gamma^{(\cdot)}$ is a *complete family of ancestors* $\mathbb{A}(\Gamma)$. A *complete family of ancestor genes* $\mathbb{G}(\Gamma)$ is a complete family of ancestors of Γ such that all members are genes.

A complete family of ancestors $\mathbb{A}(\Gamma)$ then is a family of cubic graphs which can be combined in some way, via breeding and parthenogenic operations, to precisely output Γ. Note that there may be many different complete families of ancestors for a given graph Γ. When no confusion is possible, any of them may be referred as $\mathbb{A}(\Gamma)$. However, as indicated by Theorem 3.1, we will learn that there is a unique family $\mathbb{G}(\Gamma)$ of ancestor genes of Γ.

For a given complete family of ancestors, we will choose to view an inverse operation on one of the ancestors as an inverse operation on the entire ancestral graph, but where only one of the connected components is altered. To facilitate this, we now define a generic inverse breeding operation which is to be performed on such a graph, and where the second argument makes clear which of the six breeding operations is to be performed.

Definition 3.5. Suppose Γ is a (connected or disconnected) ancestral graph, and $\Gamma_D \subseteq \Gamma$ is a connected component containing an irreducible cubic cracker C. Then, an *inverse breeding* operation on Γ is defined to be

$$\hat{\mathscr{B}}^{-1}(\Gamma, C) = \Gamma^{(\cdot)},$$

where the disconnected cubic graph $\Gamma^{(\cdot)} = (\Gamma \setminus \Gamma_D) \cup \Gamma_1 \cup \Gamma_2$ such that, Γ_D, Γ_1 and Γ_2 are as defined in Definitions 1.3, 1.4 or 1.5 if C is an irreducible 1-cracker, 2-cracker or 3-cracker respectively. Similarly, if $\Gamma_D \subseteq \Gamma$ is a connected component containing a parthenogenic object S, then an *inverse parthenogenic* operation on Γ is defined to be

$$\hat{\mathscr{B}}^{-1}(\Gamma, S) = \Gamma^{(.)},$$

where the graph $\Gamma^{(.)} = (\Gamma \setminus \Gamma_D) \cup \Gamma_1$ such that, S and Γ_1 are as defined in Definitions 1.6, 1.7 or 1.8 if S corresponds to a parthenogenic diamond, parthenogenic edge or parthenogenic triangle respectively.

For example, a sequence of inverse operations on a descendant Γ involving two inverse breeding operations could be written as $\hat{\mathscr{B}}^{-1}\big(\hat{\mathscr{B}}^{-1}(\Gamma, C_1), C_2\big) = \Gamma^{(2)}$ where C_1 is a cracker of Γ, and C_2 is a cracker of $\Gamma^{(1)} = \hat{\mathscr{B}}^{-1}(\Gamma, C_1)$. For the sake of keeping the notation simple, we will also use individual descendants (rather than ancestral graphs) as inputs for the inverse breeding operation whenever no confusion is possible.

Note that, although these definitions are given for an ancestral graph, they are applicable to be used for individual descendants by simply considering an ancestral graph containing only one descendant.

Throughout the various proofs in this chapter, it will often be necessary to show that a given set of edges do not form a cubic cracker. A technique which will often be used is to select a certain edge e from the set, and show that, despite the deletion of the specified edge set, a path still exists through the remaining edges that travels from one endpoint of e to the other. This implies that the deletion of edge e makes no contribution in disconnecting the graph. Since a cracker must be a minimal edge cut set, this technique is sufficient to prove that the edge set is not a cubic cracker.

3.2 Proof of the Main Result

The proof of Theorem 3.1 is not simple. Hence it has been broken down into a number of logical components, each discussed separately. We hope that others may discover a simpler proof of this result.

The four main stages leading to the proof of Theorem 3.1 are:

(i) Showing that there is no loss of generality in restricting our consideration to graphs that do not possess parthenogenic objects. This is established in Corollary 3.1 of Sect. 3.2.1.

(ii) Establishing a number of properties of the ancestral graph that are needed to facilitate the rest of the proofs in Sects. 3.2.2–3.2.3.

(iii) Proving that the cardinality of any complete family of ancestor genes of a graph depends only on that graph. In particular, it is invariant of the choice of the sequence of inverse operations used to obtain $\mathbb{G}(\Gamma)$. This is established in Proposition 3.9 of Sect. 3.2.3.

(iv) Defining an exposed gene of a graph Γ to be a gene that can be immediately isolated by a single inverse breeding operation, and establishing in Proposition 3.12 of Sect. 3.2.4 that if Γ is a descendant, then it possesses exposed genes.

With the help of (i)–(iv) we can then prove Theorem 3.1 by induction on the cardinality n of $\mathbb{G}(\Gamma)$.

The length of the complete proof makes it difficult to follow at times, and it can also be difficult to determine the context of some of the partial results. To aid the reader, we provide a short remark after each upcoming statement which describes the purpose of the statement and how it fits into the overall proof.

3.2.1 Ancestors of Graphs with Parthenogenic Objects

In the following discussion, we investigate the process of obtaining ancestors from a cubic graph containing at least one parthenogenic object. We will eventually show in Theorem 3.2 that for any given graph Γ containing parthenogenic objects, there exists another graph Γ_a, which is an ancestor of Γ, containing no parthenogenic objects, such that any complete family of ancestor genes for Γ is also a complete family of ancestor genes for Γ_a. This will enable us to focus purely on those graphs containing no parthenogenic objects.

Before this result is proved, we first prove three short results, Lemma 3.1, Proposition 3.1 and Lemma 3.2 which will be used to simplify certain special cases in the proof of Theorem 3.2. In that proof, the first three cases are fairly straightforward, and it is the final two cases that are aided by the three shorter results.

Consider a graph Γ with an irreducible cubic cracker C and a parthenogenic bridge S containing edge (p_1, p_2). Suppose the edges adjacent to the parthenogenic bridge are (v_1, p_1), (v_2, p_2), (u_1, p_1) and (u_2, p_2). Such a graph is displayed in Fig. 3.1.

Lemma 3.1. *The intersection of C and the edges adjacent to the parthenogenic bridge S contains at most one edge.*

Remark 3.1. This Lemma will allow us to ignore the possibility of an irreducible cubic cracker containing multiple edges adjacent to any parthenogenic object in the proof of the following proposition. This result is obvious by definition for the parthenogenic triangle, and it is impossible for an irreducible cracker to contain any edges adjacent to the parthenogenic diamond (since both such edges constitute reducible 1-crackers), so only the parthenogenic bridge needs to be considered here.

Proof. Suppose C shares at least two edges with the adjacent edges of S. Note that, by definition, no combination of edges adjacent to S forms an irreducible cubic cracker by itself. Therefore, C must be a 3-cracker containing two edges adjacent to S and one other edge. If we were to delete the parthenogenic bridge and its adjacent edges, we would isolate two connected components, say Γ_1 and Γ_2, where without loss of generality we assume that $v_1, v_2 \in V(\Gamma_1)$ and $u_1, u_2 \in V(\Gamma_2)$. The only sets of two edges adjacent to (p_1, p_2) that do not form a cutsets are $\{(v_1, p_1), (u_2, p_2)\}$ and $\{(v_2, p_2), (u_1, p_1)\}$, so by the definition of crackers we conclude that C must contain one of those two sets. First suppose $C = \{(v_1, p_1), (u_2, p_2), e\}$ where, without loss of generality, e is an edge in Γ_1. It is sufficient to show that there exists a path in Γ connecting u_2 to p_2 that does not use other edges in C, thereby contradicting the assumption that C is a cracker. Consider a path from u_2 to p_2 that does not including the edge (u_2, p_2), starting from u_2, passing some edges in Γ_2, then (u_1, p_1) and finally (p_1, p_2). Since such a path does not contain any edge of Γ_1, it does not contain e, nor does it contain (v_1, p_1). Therefore, the deletion of C does not disconnect vertices u_2 and p_2, which implies that C is not a cracker, violating the assumption. Similarly, C containing $\{(v_2, p_2), (u_1, p_1)\}$ is not a cracker. This is a contradiction, therefore C contains no more than one edge adjacent to S. $\quad\square$

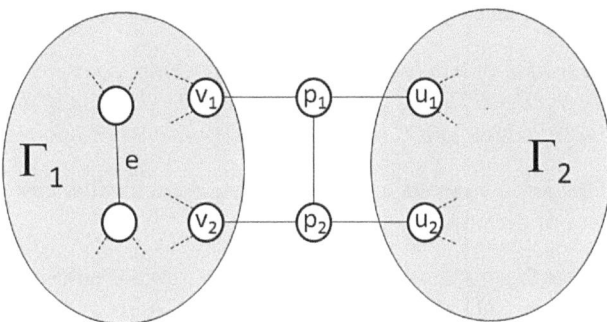

Fig. 3.1 The situation described in Lemma 3.1

Proposition 3.1. *Suppose that graph Γ contains a parthenogenic object S and an irreducible cubic cracker C. Then, at least one of the following is true:*

(i) Cracker C does not have any edge adjacent to the parthenogenic object S.
(ii) The deletion of the parthenogenic object S, its adjacent edges, and C results in exactly three connected components.

Remark 3.2. This result will be used in the proof of the upcoming Theorem 3.2 to simplify the consideration of a special case, specifically case 4, for which item (i) is not satisfied.

Proof. Since C is a cracker, its deletion results in two connected components, say Γ_1 and Γ_2. Note that, since C is irreducible, it cannot be comprised entirely of edges adjacent to S. Therefore, if S is a parthenogenic diamond or a parthenogenic triangle, then S and its remaining adjacent bridge (or bridges) lie entirely in either in Γ_1 or Γ_2. Therefore deleting S and its adjacent edges further separates the graph into three connected components, satisfying (ii).

Consider the case where S is a parthenogenic bridge. Suppose the deletion of S, its adjacent edges and C results in more than three connected components, violating (ii). It is then obvious that C is not a 1-cracker. We must show that (i) is satisfied, that is, that C does not share any edge with the adjacent edges of S. Assume the opposite is true. Then there exists an edge $e \in C$ such that e is adjacent to S. By Lemma 3.1 C does not contain any other edge adjacent to the parthenogenic bridge. Therefore, the parthenogenic bridge and all of its incident edges except e are either in Γ_1 or Γ_2. However, this implies that deletion of the parthenogenic bridge, its adjacent edges, and C results in three connected components which violates the initial assumption that (ii) is not satisfied. Therefore, C must not contain any edges adjacent to S, and so (i) is satisfied.   □

We recall from Definition 1.7, that performing a type 2 parthenogenic operation introduces a parthenogenic bridge into a 2-cracker, however the parthenogenic bridge itself is typically not a cubic cracker edge. The following lemma describes the alternative situation, where the parthenogenic bridge is a cubic cracker edge.

Lemma 3.2. *Consider an irreducible cubic cracker C that contains a parthenogenic bridge as one of its edges. Then C is a 3-cracker and the deletion of the parthenogenic bridge, its adjacent edges, and C results in four connected components.*

Remark 3.3. This lemma covers a particular case (specifically, case 5) which will arise in the proof of the upcoming Theorem 3.2.

Proof. If we delete the parthenogenic bridge (p_1, p_2) and its adjacent edges (v_1, p_1), (v_2, p_2), (u_1, p_1), (u_2, p_2), the graph is disconnected into two connected components, say Γ_1 and Γ_2, where without loss of generality we assume that $v_1, v_2 \in \Gamma_1$ and $u_1, u_2 \in \Gamma_2$. Then there exists a path between v_1 and v_2 in Γ_1 and a path between u_1 and u_2 in Γ_2.

The deletion of C separates vertices p_1 and p_2. Since u_1 must be in the same component as p_1, and u_2 is the same component as p_2, it is clear that the deletion of C must also separate u_1 and u_2. Similarly, the deletion of C separates v_1 and v_2. This implies that, in addition to (p_1, p_2), C must also have at least one edge in Γ_1 and one edge in Γ_2, and since C is a cubic cracker it therefore contains exactly three edges. The deletion of C therefore separates both Γ_1 and Γ_2 into two connected components each, resulting in a total of four connected components. This situation is displayed in Fig. 3.2.   □

Note that, although C contains a parthenogenic bridge, this situation still satisfies item (i) of Proposition 3.1, as C does not contain any edges adjacent to the parthenogenic bridge.

Theorem 3.2. *Consider a cubic graph Γ containing a parthenogenic object S. Let Γ_p be the graph obtained after performing the inverse parthenogenic operation:*

$$\Gamma_p = \hat{\mathscr{B}}^{-1}(\Gamma, S).$$

If $\Gamma^{(\cdot)}$ is an ancestral graph of Γ which is neither Γ_p, nor an ancestral graph of Γ_p, then $\Gamma^{(\cdot)}$ contains at least one parthenogenic object, and there exists either an inverse parthenogenic operation, or a sequence of two inverse parthenogenic operations that, when performed on $\Gamma^{(\cdot)}$, produces an ancestral graph of Γ_p.

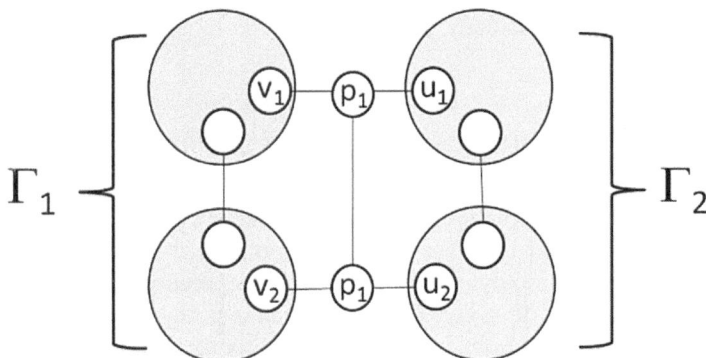

Fig. 3.2 A graph containing a 3-cracker, which in turn contains a parthenogenic bridge

Remark 3.4. The proof of Theorem 3.2 takes the form an inductive argument, where establishing the base step requires the exhaustive consideration of (five) cases that may arise. In order to facilitate the flow of this chapter, we omit the proof here. However, the full proof may be found in Appendix A.1.

We note that Γ_p in Theorem 3.2 no longer contains the parthenogenic object S. This leads to the following important corollary.

Corollary 3.1. *Any complete family of ancestor genes of a graph Γ is also a complete family of ancestor genes of graph $\Gamma_p^{(\cdot)}$ obtained after a sequence of inverse parthenogenic operations on Γ such that $\Gamma_p^{(\cdot)}$ does not contain any parthenogenic object.*

Remark 3.5. Corollary 3.1 is a vitally important result, since if we are able to prove that any cubic graph without parthenogenic objects has a unique complete family of ancestor genes, this corollary immediately implies that any cubic graph containing parthenogenic objects does as well. Then we can restrict our consideration to those graphs without parthenogenic objects, thereby greatly reducing the number of difficult situations which may arise.

Proof. From Theorem 1.1 we know that complete families of ancestor genes for Γ and Γ_p certainly exist. From Theorem 3.2 any ancestral graph of Γ that contains no parthenogenic object must either be Γ_p, or an ancestral graph of Γ_p. Inductively, this result holds not only for Γ_p, but also for any graph $\Gamma_p^{(\cdot)}$ obtained after performing a sequence of inverse parthenogenic operations on Γ_p. Then any complete family of ancestor genes of Γ is also a complete family of ancestor genes of $\Gamma_p^{(\cdot)}$. □

It should be noted that, even though Corollary 3.1 implies that we can ignore graphs with parthenogenic objects, some upcoming intermediate results will still require us to consider graphs with parthenogenic objects when establishing that certain situations cannot arise. This is because it is possible for a graph with no parthenogenic object to have ancestors that do have parthenogenic objects, and so results which seek to describe the properties of those ancestors will sometimes need to consider such cases separately.

3.2.2 Complete Family of Ancestors of Size Three

This section is devoted to the special case of obtaining a complete family of ancestors of size three. It is obvious that to obtain a complete family of ancestors of size two, we need to perform only one inverse breeding operation. Therefore the smallest case where the order of inverse operations is an important factor in obtaining the ancestor genes is when we obtain a complete family of ancestors of size three. In this section we investigate the conditions under which we obtain the same family of ancestors of size three by different sequences of operations.

Lemma 3.3. *Consider a cubic graph Γ with a complete family of ancestors $\mathbb{A}(\Gamma)$ containing n ancestors. Any sequence of inverse operations on Γ that results in the n ancestors involves exactly $n - 1$ inverse breeding operations.*

Remark 3.6. By the definition of inverse operations, Lemma 3.3 is obvious, but we formally establish it here. Note that, although two different sequences of inverse operations resulting in the n ancestors must involve the same number of inverse breeding operations, they may involve a different number of inverse parthenogenic operations.

Proof. Each inverse breeding operation increases the number of connected components by one and each inverse parthenogenic operation does not change the number of components. Since the complete family of ancestors has cardinality n, we can conclude that exactly $n - 1$ inverse breeding operations must be performed. □

As a special case, Lemma 3.3 implies that a sequence of inverse operations leading to a family of ancestors of size three involves exactly two inverse breeding operations. Hence, we focus on discussing such sequences of inverse operations.

Proposition 3.2. *Assume C_1 and C_2 are two irreducible cubic crackers of Γ which satisfy the following five properties:*

(i) *C_1 and C_2 do not share any edge.*
(ii) *If C_1 is a 1-cracker, the edges adjacent to C_1 are not contained in C_2, and vice versa.*

(iii) If both C_1 and C_2 are 1-crackers, the set of edges adjacent to C_1 is disjoint with the set of edges adjacent to C_2.

(iv) C_1 is entirely contained in one of the two subgraphs separated by C_2, and vice versa.

(v) Crackers C_1 and C_2 are irreducible in $\hat{\mathscr{B}}^{-1}(\Gamma, C_2)$ and $\hat{\mathscr{B}}^{-1}(\Gamma, C_1)$, respectively.

Then, there exists an ancestral graph $\Gamma^{(2)}$ given by

$$\Gamma^{(2)} = \hat{\mathscr{B}}^{-1}\big(\hat{\mathscr{B}}^{-1}(\Gamma, C_1), C_2\big) = \hat{\mathscr{B}}^{-1}\big(\hat{\mathscr{B}}^{-1}(\Gamma, C_2), C_1\big).$$

Remark 3.7. In this section, we consider the case where we have a complete family of ancestors of cardinality 3. Although there are quite a number of special cases, Proposition 3.2 describes the generic situation, in which there are three ancestors separated by two irreducible cubic crackers that do not interact with one another in any way, and establishes that the order in which the two crackers are removed by inverse breeding operations is inconsequential (Fig. 3.3). The special cases will be considered individually later in this section.

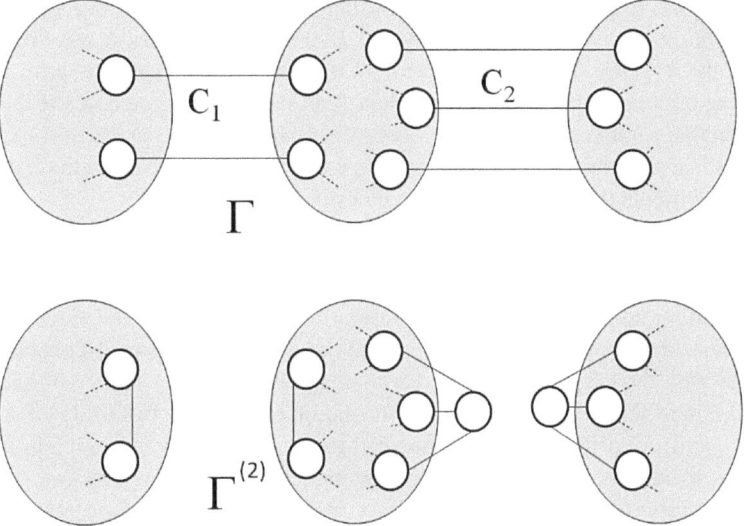

Fig. 3.3 A graph Γ containing two irreducible cubic crackers that do not share any edges, and the corresponding ancestral graph $\Gamma^{(2)}$

Proof. It is clear by the irreducibility of C_1 that $\Gamma^{(1)} = \hat{\mathscr{B}}^{-1}(\Gamma, C_1)$ exists.

Each operation involves unions and set-subtractions of sets of vertices and edges, and by assumption, the elements involved in these two operations are distinct.

Since $\hat{\mathscr{B}}^{-1}\left(\hat{\mathscr{B}}^{-1}(\Gamma,C_2),C_1\right)$ involves deleting (by set difference) and adding (by union) sets of vertices and edges which are entirely distinct, the operation is commutable. ☐

In the following definition, we introduce the concept of a *core*, which we use extensively throughout the remainder of this chapter. Roughly speaking, the cores of a graph Γ, induced by a complete family of ancestors $\mathbb{A}(\Gamma)$, contain all of the vertices and edges of Γ that remain unaltered by the sequence of inverse operations used to obtain $\mathbb{A}(\Gamma)$. Alternatively, the cores may be viewed as the elements of Γ that have been inherited directly from its ancestors, rather than introduced by the breeding operations.

Definition 3.6. Consider a complete family of ancestors $\mathbb{A}(\Gamma)$ containing n connected cubic graphs. For any graph $\Gamma_a \in \mathbb{A}(\Gamma)$, we call the subgraph $K(\Gamma_a) = \Gamma_a \cap \Gamma$ the *core* of Γ induced by Γ_a. In this case we say that Γ contains n cores induced by $\mathbb{A}(\Gamma)$.

Note that a core may be empty, in the case where all edges and vertices of an ancestor are created by inverse operations.

Definition 3.7. Consider graph Γ and a complete family of ancestors $\mathbb{A}(\Gamma)$. An inverse operation is said to *preserve cores* of Γ if no element inside any of the cores induced by $\mathbb{A}(\Gamma)$ is altered or deleted by the operation. Similarly, an irreducible cubic cracker C, or a parthenogenic object S, is said to preserve cores if the inverse operation that removes it preserves cores. A *sequence of inverse operations preserving cores* is a sequence of inverse breeding and parthenogenic operations that never involves elements that are in the induced cores.

The term "preserving cores" is chosen because we will view inverse operations as altering sections of the graph, first by deleting some edges or vertices (or both), and then by adding some in. The cores represent those segments of the graph which are never altered by the inverse operations, and hence are "preserved." This choice also indicates why the cores are defined for a particular complete family of ancestor of a graph, since the inverse operations to obtain one complete family of ancestors might be entirely different than those used to obtain a different complete family of ancestors. Note that after performing a sequence of inverse operations preserving cores, the union of those cores is a subgraph of the ancestral graph obtained at the end of the sequence.

Proposition 3.3. *Consider a graph Γ with no parthenogenic objects that contains exactly three cores induced by a complete family of ancestors $\mathbb{A}(\Gamma)$. Then one of the following is true:*

Case 1: *There are two irreducible cubic crackers C_1 and C_2 which satisfy conditions (i)–(v) of Proposition 3.2;*

Case 2: *There are two irreducible 1-crackers C_1, C_2 where C_1 and C_2 share an endpoint w. In this case the third edge connected to w, that is not in C_1 and C_2 must also be an irreducible 1-cracker;*

Case 3: *There are two irreducible 2-crackers C_1 and C_2 that share an edge e_1 and other edges in $C_1 \cup C_2$ do not share endpoints;*

Case 4: *There is a 2-cracker C_1 and a 3-cracker C_2, both irreducible, that share an edge e_1 and other edges in $C_1 \cup C_2$ do not share endpoints;*

Case 5: *There are two irreducible 3-crackers C_1 and C_2 that share an edge e_1;*

Case 6: *There is an edge $e = (b_1, b_2)$ adjacent to two irreducible 1-crackers, C_1 and C_2;*

Case 7: *There exists an edge e_1 adjacent to an irreducible 1-cracker C_1, such that e_1 is an edge of an irreducible 2-cracker C_2;*

Case 8: *There is an edge e_1 adjacent to an irreducible bridge C_1, such that e_1 is an edge of an irreducible 3-cracker C_2;*

Case 9: *There exists an inverse type 1 breeding operation applicable on Γ such that the resultant graph has a parthenogenic diamond which preserves cores induced by $\mathbb{A}(\Gamma)$;*

Case 10: *There exists an inverse type 2 breeding operation applicable on Γ such that the resultant graph has a parthenogenic diamond which preserves cores induced by $\mathbb{A}(\Gamma)$;*

Case 11: *There exists an inverse type 3 breeding operation applicable on Γ such that the resultant graph contains a parthenogenic diamond which preserves cores induced by $\mathbb{A}(\Gamma)$;*

Case 12: *There exists an inverse type 2 breeding operation applicable on Γ such that the resultant graph has a parthenogenic bridge which preserves cores induced by $\mathbb{A}(\Gamma)$;*

Case 13: *There exists an inverse type 3 breeding operation applicable on Γ such that the resultant graph has a parthenogenic bridge which preserves cores induced by $\mathbb{A}(\Gamma)$;*

Case 14: *There exists an inverse type 2 breeding operation applicable on Γ such that the resultant graph has a parthenogenic triangle which preserves cores induced by $\mathbb{A}(\Gamma)$;*

Case 15: *There exists an inverse type 3 breeding operation applicable on Γ such that the resultant graph has a parthenogenic triangle which preserves cores induced by $\mathbb{A}(\Gamma)$.*

In addition, $\hat{\mathscr{B}}^{-1}(\Gamma, C_1)$ and $\hat{\mathscr{B}}^{-1}(\Gamma, C_2)$ in Cases 1–8 and the inverse breeding operations mentioned in Cases 9–15 preserves cores of Γ induced by $\mathbb{A}(\Gamma)$.

Remark 3.8. Proposition 3.3 is important because several upcoming proofs require us to enumerate all possible situations that may arise when obtaining a complete family of ancestors of cardinality 3. These cases will be referred to again when so required in upcoming proofs in this chapter.

Proof. To establish the above we need to exhaustively inspect all possible ways that we can obtain Γ if we start with the graphs in $\mathbb{A}(\Gamma) = \{\Gamma_1^{(2)}, \Gamma_2^{(2)}, \Gamma_3^{(2)}\}$. Here we explain the steps to go through for obtaining all possible configurations and conducting the inspection.

Since $\mathbb{A}(\Gamma)$ is a complete family of ancestors, then there exists a sequence of inverse operations separating cores that outputs $\Gamma_1^{(2)}$, $\Gamma_2^{(2)}$ and $\Gamma_3^{(2)}$. Such a sequence contains exactly two inverse breeding operations because three components are obtained. Furthermore, since Γ contains no parthenogenic object, the first inverse operation must be an inverse breeding operation. Also, since the operations preserve cores, the last operation must be an inverse breeding operation because if the second inverse breeding operation were performed first, then the three components would be obtained and an inverse parthenogenic operation which preserves the cores would not be applicable.

Now, if we reverse the preceding operations we have a breeding operation followed by a sequence (possibly of zero length) of parthenogenic operations and finally another breeding operation. This sequence involves the ancestral graph $\Gamma^{(2)} = \Gamma_1^{(2)} \cup \Gamma_2^{(2)} \cup \Gamma_3^{(2)}$ and results in Γ. We will show later that this sequence of parthenogenic operations must have length one or zero.

To construct all possible configurations of Γ we must perform all possible sequences of this type and see which sequences lead to us obtaining Γ with no parthenogenic objects. The first step is to choose the first breeding operation to be performed on $\Gamma^{(2)}$. There are only three simple choices; the three breeding operations. The second step is to perform one or no parthenogenic operation (we will consider the case of performing more than one parthenogenic operation later.) Note that the parthenogenic operation must involve the newly obtained cracker because its inverse must preserve cores. Therefore, there are only a few ways to perform an inverse parthenogenic operation. At this point, we denote the obtained ancestral graph by $\Gamma^{(1)}$.

The third step is to perform another breeding operation. To investigate all cases, we must consider different edges and nodes of $\Gamma^{(1)}$ that will be deleted after the operation. Such components can be classified into three categories. Vertices and edges of the newly generated crackers and parthenogenic objects, vertices and edges adjacent to the crackers and parthenogenic objects, and all other edges. These cases are limited and it can be verified that they are all covered by one of the 15 cases considered. Note that some configurations are not valid due to the presence of parthenogenic objects in the resulting Γ. The valid configurations are illustrated in Figs. A.9–A.21. In fact, reversing the sequences of inverse operations that obtained $\Gamma^{(2)}$ from Γ in Cases 1–15, gives all possible ways to obtain a valid configuration of Γ starting from $\Gamma^{(2)}$.

Now suppose in the second step we perform a sequence of two parthenogenic operations. If we then proceed to perform step three and inspect all possible configurations as before, we observe that Γ contains at least one parthenogenic object in all cases. This implies that a breeding operation can only destroy one parthenogenic object. Therefore, if we perform more parthenogenic operations in step two we do not obtain a valid configuration. This implies that the considered 15 cases cover all possible configurations. $\square$

In the above proof we inspected the case where a breeding operation was performed on a graph with two or more parthenogenic objects and we demonstrate that the resultant graph always contains at least one parthenogenic object. This finding is formulated in the corollary below.

Corollary 3.2. *Consider a graph Γ with no parthenogenic objects. Suppose an inverse breeding operation is performed on Γ resulting in the ancestral graph $\Gamma^{(1)}$ and $\Gamma_a^{(1)}$ is a components of $\Gamma^{(1)}$. Then $\Gamma_a^{(1)}$ contains at most one parthenogenic object.*

Remark 3.9. This incidental result turns out to be useful later in this chapter when handling a special case.

Proposition 3.4. *Consider a graph Γ which contains exactly three cores induced by a complete family of ancestors $\mathbb{A}(\Gamma)$. Then suppose some sequence of inverse operations is performed such that each inverse operation preserves cores induced by $\mathbb{A}(\Gamma)$. Furthermore, after performing this sequence of inverse operations, it is no longer possible to perform any inverse operations that preserve cores induced by $\mathbb{A}(\Gamma)$. Then,*

(i) the cores induced by $\mathbb{A}(\Gamma)$ have been separated, and
(ii) the resulting ancestral graph is precisely the ancestral graph associated with $\mathbb{A}(\Gamma)$.

Remark 3.10. Suppose that, for a given graph, a complete family of ancestors of cardinality 3 is identified. Proposition 3.4 establishes that any sequence of inverse operations which preserves the cores corresponding to those three ancestors, gives a unique result. Later in this chapter, it will be established that if the three ancestors are all genes, any sequence of inverse operations preserves those cores, establishing the uniqueness result for the case when the complete family of ancestor genes has cardinality bounded above by 3. This, in turn, will form the base case for an inductive argument in the proof of the main result.

In order to help maintain the logical flow of this chapter, we omit the proof of Proposition 3.4 here as it requires consideration of all 15 cases described in Proposition 3.3. However, the full proof can be found in Appendix A.2.

Considering all possible inverse breeding operations on all configurations in the proof of Proposition 3.4 leads to the corollary below.

Corollary 3.3. *Consider graph Γ and a complete family of ancestors $\mathbb{A}(\Gamma) = \{\Gamma_1, \Gamma_2, \Gamma_3\}$. Then one of the following is true:*

(i) There are two irreducible 3-crackers C_1 and C_2 that preserve cores induced by $\mathbb{A}(\Gamma)$ that share an edge e_1. In this case, two of the three ancestors in $\mathbb{A}(\Gamma)$ can be obtained directly by an inverse breeding operation, $\hat{\mathscr{B}}^{-1}(\Gamma, C_1)$ and $\hat{\mathscr{B}}^{-1}(\Gamma, C_2)$ respectively.
(ii) There is an edge e adjacent to two irreducible 1-crackers, C_1 and C_2, that preserve cores induced by $\mathbb{A}(\Gamma)$. In this case, two of the three ancestors in $\mathbb{A}(\Gamma)$ can be obtained directly by an inverse breeding operation, $\hat{\mathscr{B}}^{-1}(\Gamma, C_1)$ and $\hat{\mathscr{B}}^{-1}(\Gamma, C_2)$ respectively.

(iii) For each of the ancestors in $\mathbb{A}(\Gamma)$, there exists an inverse breeding operation that can isolate the ancestor (as a component of the resultant ancestral graph).

Remark 3.11. Note that items (i) and (ii) of Corollary 3.3 correspond to cases 5 and 6 in Proposition 3.3 and are considered in the proof of Proposition 3.4 included in Appendix A.2. Item (iii) covers all other situations. This corollary will be used when proving a later result, where it will be established that, subject to certain conditions, it is always possible to immediately obtain an ancestor with desirable properties by performing a single inverse breeding operation.

Proposition 3.5. *Consider graph Γ containing two irreducible cubic crackers C_1 and C_2 such that one of the subgraphs of Γ separated by C_1 does not contain any edges of C_2, and vice versa. Let Γ_1 be the connected component of $\Gamma_1^{(1)} = \mathscr{B}^{-1}(\Gamma, C_1)$ that does not contain any edges originally belonging to C_2. Then,*

(i) Γ_1 can be obtained by performing an inverse breeding operation, or by performing an inverse parthenogenic operation followed by an inverse breeding operation, on the graph $\Gamma_2^{(1)} = \mathscr{B}^{-1}(\Gamma, C_2)$.

(ii) Also, if Γ_2 is the of $\Gamma_2^{(1)}$ that does not contain any edges originally belonging to C_1, there exists a unique graph Γ_3 such that $\{\Gamma_1, \Gamma_2, \Gamma_3\}$ is a complete family of ancestors of Γ.

Remark 3.12. The conditions of this proposition have been deliberately formulated so that the possible cases that arise may be confirmed to correspond precisely to those listed in Proposition 3.3.

Proof. In order to prove this result, we must consider all possible configurations of two irreducible cubic crackers. By inspection, it can be easily confirmed that these configurations are precisely those identified in Proposition 3.3 and then considered in the proof of Proposition 3.4. In order to identify all cases, one should consider all possible ways for two crackers to share edges and endpoints. For the cases where one of the crackers, say C_1, is a 1-cracker, the cases in which C_2 contains an edge that is adjacent to C_1 should also be considered separately. Then following each case from the proof of Proposition 3.4, it can be quickly confirmed that (i) and (ii) are satisfied in all possible cases. $\quad\square$

3.2.3 Sequence of Inverse Breeding Operations

In this section, we discuss some of the properties of the ancestral graph $\Gamma^{(\cdot)}$ obtained after a sequence of inverse operations on Γ. We first investigate how the cracker and non-cracker edges in $\Gamma^{(\cdot)}$ relate to the cracker and non-cracker edges in Γ. These results allow us to show that if the cardinality of any one complete family of ancestor genes of Γ is n, then any complete family of ancestor genes which results from a sequence of inverse operations on Γ also has cardinality n.

Lemma 3.4. *Consider a connected cubic graph Γ containing at least two cubic crackers C_1 and C_2. Suppose both subgraphs separated by C_1 contain edges of C_2, and vice versa. Then precisely one of the following must be satisfied:*

(i) *The deletion of C_1 and C_2 from Γ results in a graph with exactly four connected components.*

(ii) *Both C_1 and C_2 are 3-crackers and the deletion of C_1 and C_2 from Γ results in a graph with exactly five connected components.*

Remark 3.13. This result will be used to simplify one case (specifically, case 1.2) in the next proposition. It should be noted that C_1 and C_2 being 3-crackers is a necessary but not sufficient condition for their removal to result in a graph with five connected components.

Proof. Consider the subgraph Γ_s obtained by deleting C_1 from Γ. Since crackers are minimal cutsets, it is clear that this subgraph has two connected components, say $\Gamma_1(C_1)$ and $\Gamma_2(C_1)$. By the assumption of this lemma, we assume that both $\Gamma_1(C_1)$ and $\Gamma_2(C_1)$ contain at least one edge of C_2. Now, it might be the case that C_1 and C_2 share an edge, however they cannot share multiple edges, or else it would be impossible for an edge of C_2 to remain in both $\Gamma_1(C_1)$ and $\Gamma_1(C_2)$ since C_2 is a cubic cracker and has at most 3 edges. It is clear that the remaining (possibly all) edges of C_2 in Γ_s still form a cutset, since C_2 was a minimal cutset in Γ. Then if we delete the remaining edges of C_2 from Γ_s, both $\Gamma_1(C_1)$ and $\Gamma_2(C_1)$ must be further disconnected. Therefore there will be at least four connected components once both crackers are deleted. Since, by the definition of crackers, the removal of C_1 must separate the graph into precisely two connected components, the only way it is possible to end up with five connected components after deleting both C_1 and C_2 is if C_2 is a 3-cracker, and it so happens that the deletion of each edge of C_2 increases the number of connected components by 1. An equivalent argument can be used to show that C_1 must also be a 3-cracker for this to be possible, completing the proof. $\square$

Proposition 3.6. *Consider a connected cubic graph Γ which is a descendant. Consider an edge e which is an element of a cubic cracker C_1 in Γ. Then:*

(i) *If an inverse breeding or inverse parthenogenic operation $\hat{\mathscr{B}}^{-1}(.)$ is performed on Γ resulting in $\Gamma^{(1)}$ such that $e \in E(\Gamma^{(1)})$, then e is also a cubic cracker edge in $\Gamma^{(1)}$.*

(ii) *Suppose, in addition to (i), that both $\hat{\mathscr{B}}^{-1}(.)$ and C_1 preserve cores of Γ induced by a complete family of ancestors $\mathbb{A}(\Gamma)$. Then, there exists a cubic cracker $C^{(1)}$ in $\Gamma^{(1)}$ such that $e \in C^{(1)}$ and $C^{(1)}$ preserves cores induced by $\mathbb{A}(\Gamma)$ in $\Gamma^{(1)}$.*

Remark 3.14. Proposition 3.6 establishes that any cubic cracker edge which is not removed or altered by an inverse operation, remains a cubic cracker edge in the subsequent ancestral graph, and so it is impossible for an inverse operation to change the nature of a remaining cubic cracker edge. Furthermore, the cracker in the new ancestral graph which contains the edge will preserve the same cores as the original cracker C_1.

Proof. Proof of Proposition 3.6 requires arguments very similar to the ones we have used earlier in this chapter. Therefore we only present an outline of the proof. In this proof we consider performing inverse breeding and inverse parthenogenic operations in different situations and verifying the proposition in each case, using the same technique as used in the proofs of Proposition 3.4 and Theorem 3.2.

1. *Inverse breeding operations*

Note that Since graph $\Gamma^{(1)}$ still contains e, the inverse operation $\hat{\mathscr{B}}^{-1}$ is not $\hat{\mathscr{B}}^{-1}(\Gamma, C_1)$. Therefore the inverse breeding operation $\hat{\mathscr{B}}^{-1}$ involves an irreducible cracker $C_2 \neq C_1$ in Γ. There are a number of subcases to consider. To prove item (i) we show that in every case there is a cracker that contains e in $\Gamma^{(1)}$. To prove item (ii) we consider the subgraph Γ_s of Γ obtained by deleting C_1 and C_2 in each subcase, and the edges adjacent to bridges in $C_1 \cup C_2$ (if either or both are 1-crackers). From the definition of cores, it is clear that Γ_s contains the cores induced by $\mathbb{A}(\Gamma)$. Then, all we need to show is that in each case, a cubic cracker $C^{(1)}$ exists in $\Gamma^{(1)}$, and it connects components of Γ_s.

Case 1.1: Cracker C_1 does not have edges in both subgraphs isolated by the deletion of C_2, and vice versa. Then, as argued previously in the proof of Proposition 3.5, it can be seen that the cases we have to consider are same as those identified in Proposition 3.3 and then considered in the proof of Proposition 3.4. In each case we have to go through all choices of C_2. Then, after performing the operation $\hat{\mathscr{B}}^{-1}(\Gamma, C_2)$ to obtain $\Gamma^{(1)}$, we can show that for all valid choices of e (where e is an edge in a cracker C_1 in Γ and $e \in E(\Gamma^{(1)})$), there exists a cubic cracker $C^{(1)}$ in $\Gamma^{(1)}$ that contains e and also connects components of Γ_s.

Case 1.2: Cracker C_1 contains an edge in both subgraphs separated by C_2. By Lemma 3.4 this implies that Γ_s has four or five components. We use proof by contradiction to show that edges of C_1 and C_2 are non-adjacent (otherwise C_1 and C_2 are not crackers), and neither can be a 1-cracker (otherwise Γ_s has three components). Then, using the same procedure as in Case 1.1 we go through each case to show that edges in C_1 are still cracker edges of new cubic crackers in $\Gamma^{(1)}$ that contain edges introduced by an inverse breeding operation. The possible cases are:

 (i) Crackers C_1 and C_2 are 2-crackers.
 (ii) Crackers C_1 and C_2 are 3-crackers and their deletion results in 4 components.
 (iii) Cracker C_1 is a 2-cracker and cracker C_2 is a 3-cracker.
 (iv) Cracker C_1 is a 3-cracker and cracker C_2 is a 2-cracker.
 (v) Crackers C_1 and C_2 are 3-crackers and their deletion results in 5 components.

Note that in the cases above, we can show that $C_1 \cap C_2 = \emptyset$, otherwise Γ_s would have three components and is considered in Case 1.1.

The above cases are illustrated in Fig. 3.4.

2. *Inverse parthenogenic operations*

There are three cases to consider. To prove item (i) we show that in every case there is a cracker that contains e in $\Gamma^{(1)}$. To prove item (ii) we consider the subgraph Γ_s of Γ obtained by deleting C_1 in each case, those adjacent edges to C_1 that are bridges, and a parthenogenic object S and its adjacent edges. Note that Γ_s is subgraph of $\Gamma^{(1)}$ as well. Also if $\hat{\mathscr{B}}^{-1}$ is an inverse operation preserving cores and C_1 separates the induced cores, then Γ_s certainly contains all the cores induced by $\mathbb{A}(\Gamma)$. Therefore all we need to show is that in each case, a cracker $C^{(1)}$ exists and connects components of Γ_s.

Case 2.1: C_1 does not share any edge with those adjacent to S, or any edges of S. We can use the argument used in Case 1 of the proof of Theorem 3.2 to show that C_1 remains a cracker in $\Gamma^{(1)}$ and connected components of Γ_s.

Case 2.2: C_1 contains at least one edge adjacent to S. First we need to show that S can only be a parthenogenic bridge or a parthenogenic triangle. To do so, assume S is a parthenogenic diamond. It is straight forward to show that two adjacent edges to S are both bridges, hence C_1 must be a 1-cracker and e must be one of the bridges adjacent to S. However, in this case $\hat{\mathscr{B}}^{-1}(\Gamma, S)$ would delete e, which violates the assumption. Therefore S is not a parthenogenic diamond. As a result S is either a parthenogenic bridge or parthenogenic triangle. By the same argument, C is either a 2-cracker or a 3-cracker. This gives us the four possible situations

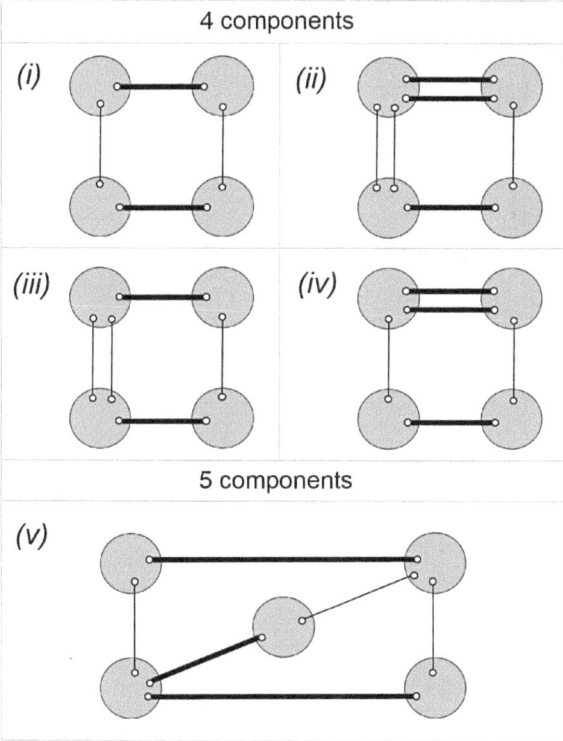

Fig. 3.4 Five subcases of Case 1.2 of Proposition 3.6. Crackers C_1 and C_2 are displayed by *bold lines* and *thin lines* respectively

displayed in Fig. 3.5. We can use an argument similar to the one used in Case 4 of the proof of Theorem 3.2 to show that $\Gamma^{(1)}$ contains an edge introduced by the inverse parthenogenic operation $\hat{\mathscr{B}}^{-1}$.

Case 2.3: C_1 contains an edge of S. Since disconnecting any two vertices of a parthenogenic diamond or triangle requires the deletion of adjacent edges, it is clear that none of those edges can be part of a cracker, and so S must be a parthenogenic bridge. Furthermore, by Lemma 3.2, C_1 must be a 3-cracker. This case is similar to Case 5 in the proof of Proposition 3.2.

The above situations constitute all possible situations, completing the proof. □

Corollary 3.4. *The cores of graph Γ induced by a complete family of ancestor genes $\mathbb{G}(\Gamma)$ do not contain any cubic crackers edges.*

Remark 3.15. Although it follows very swiftly from Proposition 3.6, this corollary (in conjunction with another upcoming corollary) will be vital in proving that, if it were the case that a graph had multiple complete families of ancestor genes, that they would all have the same cardinality.

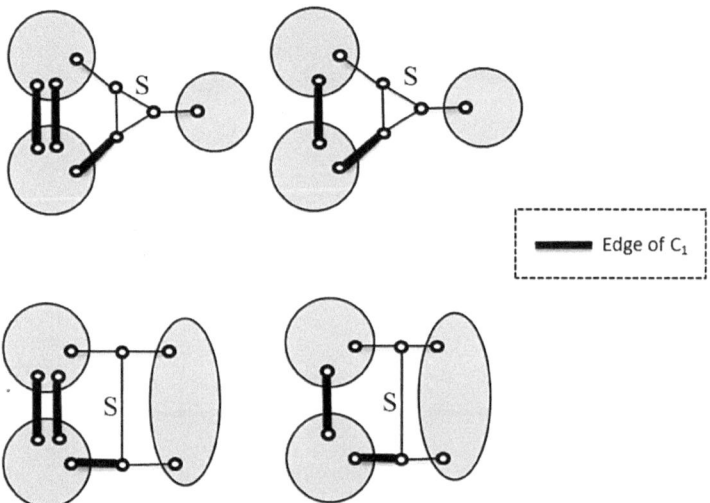

Fig. 3.5 The possible situations in Case 2.2 of Proposition 3.6 where C_1 contains at least one edge adjacent to the parthenogenic object S

Proof. Suppose the opposite is true. Then, there exists a cubic cracker edge e contained in a core induced by $\mathbb{G}(\Gamma)$. Consider a sequence of inverse operations that obtain the ancestral graph $\Gamma^{(\cdot)}$ such the family of all connected components in $\Gamma^{(\cdot)}$ is $\mathbb{G}(\Gamma)$. Since e is contained in an induced core it is obvious, by definition, that e is also in $\Gamma^{(\cdot)}$. Since connected components of $\Gamma^{(\cdot)}$ are genes, then e is not in a cubic cracker in $\Gamma^{(\cdot)}$. Therefore when performing the sequence of inverse operations on Γ, at some point e is a cubic cracker edge, but after one further inverse operation it is no longer a cubic cracker edge. However, by Proposition 3.6 this is a contradiction. Therefore, the corollary holds. □

Of course, cores of a graph Γ induced by a complete family of ancestors may still contain cubic cracker edges if not all of the complete family of ancestors are genes.

Proposition 3.7. *Consider Γ with n cores induced by a particular complete family of ancestors $\mathbb{A}(\Gamma)$. The number of inverse breeding operations, that preserve cores, required to separate all of the induced cores is $n - 1$.*

Remark 3.16. After any sequence of inverse operations preserving cores is performed, two cases may arise. Either is possible to perform additional inverse operations preserving cores in the resultant ancestral graph, or it is not possible to do so. The latter case corresponds precisely to the situation when all of the induced cores have been separated, and this proposition states that any sequence of the latter type must contain exactly $n - 1$ inverse breeding operations. This result will be used in the proof that the cardinality of any complete family of ancestor genes for a given graph is an invariant of that graph.

Proof. By Proposition 3.6, after each inverse operation performed on Γ that preserve cores, any cubic crackers in the resultant graph that share no elements with the cores induced by $\mathbb{A}(\Gamma)$, must connect those cores. Therefore, another inverse operation is applicable that preserves the induced cores. Inductively, we can continue this process until all induced cores are separated. Since every inverse breeding operation increases the number of connected components by one, and every inverse parthenogenic operation does not change the number of connected components, the number of inverse breeding operations preserving cores required to separate all induced cores is $n - 1$. □

Note that in the following proposition we consider (standard) breeding and parthenogenic operations, rather than inverse breeding and inverse parthenogenic operations.

Proposition 3.8. *Consider an edge e of cubic cracker C in a cubic graph Γ_a. Suppose a breeding or parthenogenic operation is performed involving Γ_a and results in Γ such that e is also in Γ. Then e is a cubic cracker edge in Γ as well.*

Remark 3.17. Proposition 3.8 establishes a similar result to that of Proposition 3.6, except in the opposite direction. Effectively, the proposition establishes that any cubic cracker edges in a parent graph which are still present in the descendant graph are cubic cracker edges there as well. More importantly, this result implies that any edges in a descendant which are not cubic cracker edges, but are present in a parent graph, are not cubic cracker edges in the parent graph either.

Proof. It is clear that the breeding or parthenogenic operation must not involve the edge e, or else it will not be present in Γ. Then there are two cases to consider. First, consider a breeding or parthenogenic operation that does not involve any edges of C. Suppose C separates Γ_a into two subgraphs Γ_a^1 and Γ_a^2. Then, since the breeding or parthenogenic operation does not involve any edges of C, it is clear that the elements of Γ_a which are involved in the operation are all contained in either Γ_a^1, or Γ_a^2. In this case, C still separates two subgraphs of Γ because there will be no other connections between the two subgraphs, and since it is unaltered, it remains a cubic cracker.

The second case to consider is when an edge e_1 of C, that is not e, is involved in the operation resulting in Γ. Obviously C is not a bridge. Also, since by definition no edge of C can be a bridge, then type 1 and type 3 parthenogenic operations are not applicable. The possible cases are illustrated in Fig. 3.6. In the cases of types 1 and 2 breeding, the operation involves edge e_1 being deleted. In the case of type 3 breeding, the operation involves an endpoint of edge e_1 being deleted. In the case of type 2 parthenogenesis, the operation involves a 2-cracker containing edge e_1 being deleted. In each case it can be checked that e belongs to a cubic cracker in the resultant graph Γ.   □

Corollary 3.5. *Consider an edge $e \in \Gamma$, such that no cubic cracker in Γ contains e. Suppose an inverse operation is performed on Γ resulting in an ancestral graph $\Gamma^{(1)}$. Then, there is no cubic cracker in $\Gamma^{(1)}$ that contains e.*

Remark 3.18. This corollary, along with Corollary 3.4, establishes that it is impossible to perform an inverse operation and, in doing so, change a cubic cracker edge into a non-cubic cracker edge, or vice versa. These two results will be used in the upcoming Proposition 3.9.

Proof. Suppose the opposite is true. Then, e is a cubic cracker edge in $\Gamma^{(1)}$. Since Γ can be constructed via a breeding or a parthenogenic operation involving connected components of $\Gamma^{(1)}$, and Γ still contains edge e, then by Proposition 3.8, e is a cubic cracker edge in Γ as well, which is a contradiction.   □

For the following proposition, we look at the cardinality of a complete family of ancestor genes.

Proposition 3.9. *The number of inverse breeding operations required to obtain any complete family of ancestor genes is fixed for a given graph, or equivalently, any complete family of ancestor genes in Γ has fixed cardinality depending only on the graph.*

Remark 3.19. Although, ultimately, we will conclude that every cubic graph has a unique complete family of ancestor genes, at this intermediate stage it is still conceivable that there might be several different complete families of ancestor genes for a given graph. Proposition 3.9 implies that every one of these complete families of ancestor genes has the same cardinality, with that cardinality then being an invariant of the graph. This restriction will allow us to use the cardinality as the subject of an inductive argument to eventually conclude that only a unique complete family of ancestor genes exists for any given graph.

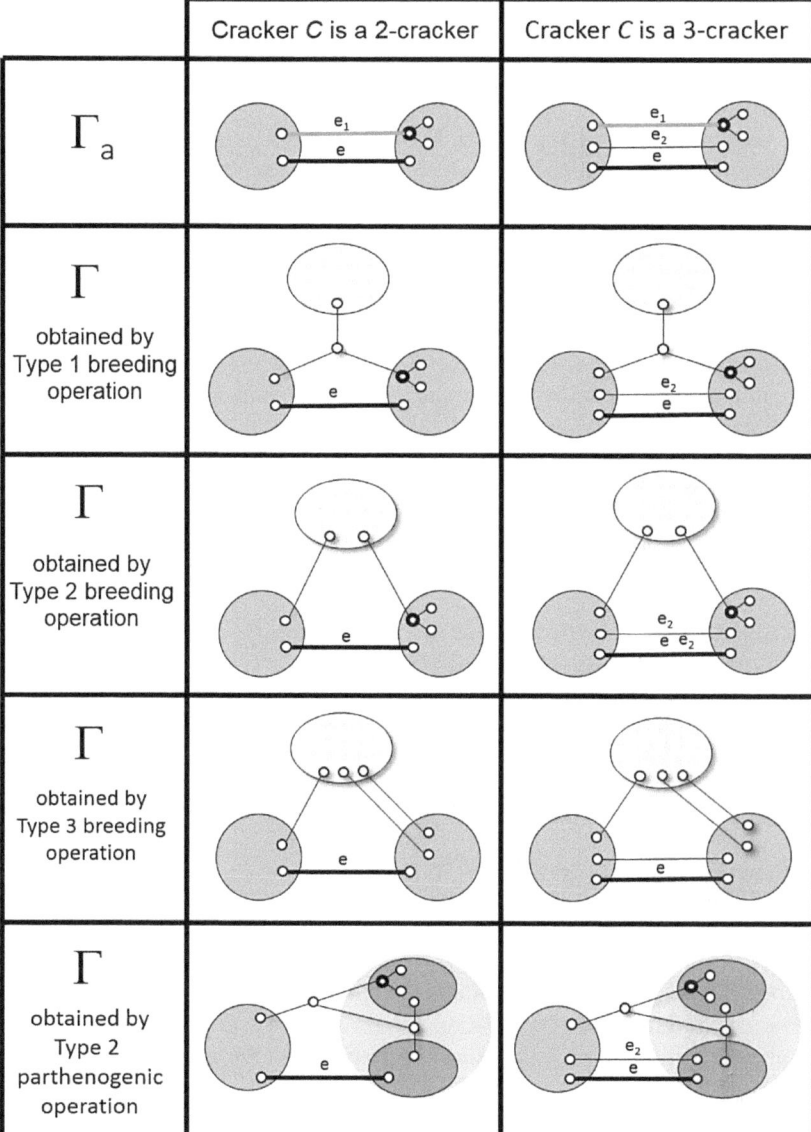

Fig. 3.6 The possible cases in which a cracker edge e_1 of C is deleted as a result of a breeding or parthenogenic operation

Proof. From Theorem 1.1 it is always possible to obtain a complete family of ancestor genes, say $\mathbb{G}(\Gamma)$, by simply performing any valid sequence of inverse operations until no application inverse operations remain. Suppose that for this complete family of ancestor genes, the cardinality is n. We will first show that any sequence of inverse operations on Γ which results in a complete family of ancestor genes (possibly different from $\mathbb{G}(\Gamma)$) is a sequence of inverse operations that preserves cores induced by $\mathbb{G}(\Gamma)$.

Any inverse breeding operation performed on Γ preserves cores induced by $\mathbb{G}(\Gamma)$, because by Corollary 3.4 we know that cores induced by a family of ancestor genes contain no cubic crackers. Furthermore, by Corollary 3.5, such an inverse breeding operation cannot create a new cubic cracker C_p in the resultant graph in such a way that C_p contains an edge inside a core induced by $\mathbb{G}(\Gamma)$. Therefore the union of cores induced by $\mathbb{G}(\Gamma)$ is always a subgraph of any ancestral graph obtained after a sequence of inverse operations on Γ. Also, by Proposition 1.4, a sequence of inverse operations of sufficient length will always result in a complete family of ancestor genes. Therefore any sequence of inverse operations on Γ that results in a complete family of ancestor genes is a sequence of inverse operations that preserves cores induced by $\mathbb{G}(\Gamma)$. By Proposition 3.7, the number of inverse breeding operations in such a sequence is $n - 1$, and therefore it is a fixed constant for a given graph.   □

Since, from Proposition 3.9, we know that the cardinality of any complete family of ancestor genes is fixed for a given graph, the following definition is unambiguous.

Definition 3.8. The *number of genes* in a cubic graph Γ is the (fixed) cardinality of any of its complete families of ancestor genes.

For the sake of simplicity, in graphs where the number of genes is k, we will say the graph *has k genes*. In the case where Γ is itself a gene, by default we say that Γ has one gene.

The next corollary follows immediately from Propositions 3.4 and 3.9.

Corollary 3.6. *Consider a cubic graph Γ, with a complete family of ancestor genes $\mathbb{G}(\Gamma)$ such that $|\mathbb{G}(\Gamma)| = 3$. Then $\mathbb{G}(\Gamma)$ is the unique complete family of ancestor genes for Γ.*

Remark 3.20. Now that we have established that the number of genes for a graph is an invariant of that graph, Proposition 3.4 constitutes a proof of the uniqueness of the complete family of ancestor genes for graphs with three genes.

Proof. Suppose there exist another complete family of ancestor genes $\mathbb{G}_2(\Gamma)$. By Proposition 3.9, $|\mathbb{G}_2(\Gamma)| = 3$. From the proof of Proposition 3.9, we know that any sequence of inverse operations on Γ preserves cores induced by $\mathbb{G}(\Gamma)$. Therefore the cores induced by $\mathbb{G}_2(\Gamma)$ are the same as those induced by $\mathbb{G}(\Gamma)$. Then, by Proposition 3.4 we know that any sequence of inverse operations that preserves the three cores induced by $\mathbb{G}(\Gamma)$ results in a unique complete family of ancestors; in this case, a unique complete family of ancestor genes.   □

Of course, Corollary 3.6 also implies that the union of the genes of any complete family of ancestors of Γ with three genes is also $\mathbb{G}(\Gamma)$.

3.2.4 Existence of Exposed Genes

In this section, we show that for any given graph Γ with no parthenogenic objects, there exists at least one inverse breeding operation that separates one gene. This result is required in the next section when proving the uniqueness of the complete family of ancestor genes.

Definition 3.9. If an inverse breeding operation can be performed on a descendant graph Γ, such that one of the two components of the resultant ancestral graph is a gene, then that gene is called an *exposed gene* in Γ. Similarly, given a complete family of ancestors $\mathbb{A}(\Gamma)$, if an inverse breeding operation on Γ outputs an ancestor in $\mathbb{A}(\Gamma)$, the core corresponding to that ancestor is called an *exposed core* in Γ.

Lemma 3.5. *Consider Γ and an exposed gene G in Γ. Then:*

(i) *If G is obtained after an inverse type 1 breeding operation involving a 1-cracker $C = \{e\}$, then the only connections of $K(G)$, the core induced by G, to the rest of the graph are by two edges, each connected to one of the endpoints of e.*

(ii) *If G is obtained after an inverse type 2 or type 3 breeding operation involving a cracker C, then the only connections of $K(G)$ to the rest of the graph are the edges of C.*

Remark 3.21. This lemma describes how a cracker whose removal isolates a gene must relate to that gene. This result is useful later when determining how the isolation of an exposed gene can affect the nature of any other exposed genes in the same graph.

Proof. It follows from the definition of $K(G)$ that $K(G) = G \cap \Gamma$. Since G is an exposed gene, it can be isolated by a single inverse breeding operation. It follows from the definitions of the inverse breeding operations, and of cores, that whenever a graph, G in this case, can be isolated from Γ by a single inverse breeding operation, $K(G)$ is connected to the rest of Γ as described in items (i) and (ii). $\square$

Lemma 3.6. *Consider a graph Γ with no parthenogenic objects constructed via a breeding operation $\mathcal{B}$ on two graphs $\Gamma_a^{(1)}$ and $\Gamma_b^{(1)}$. Suppose G is an exposed gene of $\Gamma_a^{(1)}$. Then G is also an exposed gene in Γ if the breeding operation $\mathcal{B}$ does not involve the following:*

(i) *any elements (edges and vertices) of $K(G)$ in $\Gamma_a^{(1)}$, or*
(ii) *any edges adjacent to $K(G)$ that are adjacent to a bridge.*

Remark 3.22. This lemma investigates the ways in which an exposed gene may become no longer exposed after a breeding operation is performed. It will later be seen that for any graph with multiple exposed genes, it is only possible for a breeding operation to involve elements listed in items (i) or (ii) for a maximum of one of the exposed genes, ensuring that the rest remain exposed.

Proof. Since G is an exposed gene in $\Gamma_a^{(1)}$, then there exists $\Gamma_a^{(2)}$ such that an inverse breeding operation on $\Gamma_a^{(1)}$ outputs the ancestral graph $G \cup \Gamma_a^{(2)}$. So $\{G, \Gamma_a^{(2)}\}$ is a complete family of ancestors for $\Gamma_a^{(1)}$. This implies that $\{G, \Gamma_a^{(2)}, \Gamma_b^{(1)}\}$ is a complete family of ancestors for Γ. By Corollary 3.3, G is guaranteed to be an exposed gene of Γ except in two cases below

(i) There are two irreducible 3-crackers C_1 and C_2 that separate cores induced by $\mathbb{A}(\Gamma)$ that share an edge e_1. In this case, two of the three ancestors in $\mathbb{A}(\Gamma)$ can be obtained directly by an inverse breeding operation ($\hat{\mathscr{B}}^{-1}(\Gamma, C_1)$ and $\hat{\mathscr{B}}^{-1}(\Gamma, C_2)$).

(ii) There is an edge e adjacent to two irreducible 1-crackers, C_1 and C_2, that separate cores induced by $\mathbb{A}(\Gamma)$. In this case, two of the three ancestors in $\mathbb{A}(\Gamma)$ can be obtained directly by an inverse breeding operation ($\hat{\mathscr{B}}^{-1}(\Gamma, C_1)$ and $\hat{\mathscr{B}}^{-1}(\Gamma, C_2)$).

Note that the two cases above correspond to the Cases 5 and 6 in Proposition 3.3 and then considered in the proof of Proposition 3.4 in Appendix A. Then we can inspect how these two cases might arise by simply reversing the sequence of inverse operations that were considered in Proposition 3.4, and taking G as one of the ancestors in $\mathbb{A}(\Gamma)$. Performing this check in both cases reveals that, for the core $K(G)$ in $\Gamma^{(1)}$ to be unexposed in Γ, it must be the case that the breeding operation involves one of elements listed in items (i) and (ii), hence violating the condition of the lemma. Therefore we conclude that in all cases satisfying the stated conditions, G is exposed in Γ. $\square$

Corollary 3.7. *Consider Γ with two exposed genes G_1 and G_2. Then there exists a unique graph $\Gamma_a^{(1)}$ such that $\{G_1, G_2, \Gamma_a^{(1)}\}$ is a complete family of ancestors of Γ.*

Remark 3.23. This result is a special case of Proposition 3.5 and is given here without proof.

Lemma 3.7. *Consider $\Gamma_a^{(1)}$, a graph with two exposed genes G_1 and G_2. Suppose a breeding operation $\mathscr{B}$ on $\Gamma_a^{(1)}$ and another graph, say $\Gamma_b^{(1)}$, outputs Γ with no parthenogenic objects. Then G_2 is an exposed gene in Γ if $\mathscr{B}$ involves either of the following:*

(i) any elements (edge and vertex) of the $K(G_1)$ in $\Gamma_a^{(1)}$,
(ii) any edges adjacent to $K(G_1)$ that is adjacent to a bridge.

Remark 3.24. This result will be used shortly to prove that a single breeding operations can only prevent one gene at a time from remaining exposed in the resultant graph.

Proof. By Corollary 3.7, there exists $\Gamma_a^{(2)}$ such that $\mathbb{A}(\Gamma) = \{G_1, G_2, \Gamma_a^{(2)}\}$ is a complete family of ancestors of $\Gamma_a^{(1)}$. Now consider the cores $K(G_1)$ and $K(G_2)$ induced by $\mathbb{A}(\Gamma)$ in $\Gamma_a^{(1)}$. In order to prove that G_2 is an exposed gene in Γ, we will demonstrate that $\mathscr{B}$ satisfies the conditions of Lemma 3.6. Specifically, we will demonstrate that $\mathscr{B}$ does not involve the following:

(a) an element (edge and vertex) of $K(G_2)$ in $\Gamma_a^{(1)}$,

(b) an edges adjacent to $K(G_2)$ that is adjacent to a bridge in $\Gamma_a^{(1)}$.

In order to investigate the above, let us recall the elements deleted by a breeding operation. In case of type 1 and type 2 breeding, only a single edge is deleted, and in the case of type 3 breeding a vertex is deleted (and obviously its incident edges are removed as well).

Consider the case where $\mathscr{B}$ removes a single edge e. Suppose e is one of the elements listed in condition (i). Since $K(G_1)$ and $K(G_2)$ are distinct it is obvious that the e can not also be one of the elements listed in (a). Furthermore, $K(G)$ can not include edges adjacent to a bridge endpoint because it is a core induced by a gene, so e cannot be one of the elements listed in (b) either. A similar argument can be made if e is an element listed in condition (ii).

Next, consider the alternative case when $\mathscr{B}$ removes a vertex v and its incident edges e_1, e_2 and e_3. Note that in this case $\mathscr{B}$ must be a type 3 breeding operation. Obviously e_1, e_2 and e_3 are not in $K(G_2)$ since their endpoint v is not in $K(G_2)$, so any elements listed in (a) will not be removed by $\mathscr{B}$. Furthermore, since v can neither be a bridge endpoint (due to the definition of type 3 breeding), nor a vertex in $K(G_2)$, then none of e_1, e_2 and e_3 are edges listed in (b).

We have shown that in all possible cases, $\mathscr{B}$ can not involve any elements described in (a) or (b), and therefore by Lemma 3.6, G_2 is an exposed gene in Γ. $\square$

Proposition 3.10. *Consider* $\Gamma_a^{(1)}$ *with at least two exposed genes. If* $\Gamma_a^{(1)}$ *breeds with another graph, say* $\Gamma_b^{(1)}$ *resulting in* Γ *with no parthenogenic objects, then at least one of the exposed genes of* $\Gamma_a^{(1)}$ *is an exposed gene of* Γ.

Remark 3.25. This result implies that when any descendant containing at least two exposed genes is involved in a breeding (not parthenogenic) operation, at least one of those gene remains exposed in the child graph. Since breeding operations always involve two graphs, if both parents have at least two exposed genes then the child graph does as well. This result will be used as the inductive step of a proof that all descendants without parthenogenic objects have at least two exposed genes.

Proof. Let G_1 and G_2 be two exposed gene of $\Gamma_a^{(1)}$. Then there exists an inverse breeding operations on $\Gamma_a^{(1)}$ that outputs $\Gamma^{(2)} = G_1 \cup \Gamma_a^{(2)}$ where G_1 and $\Gamma_a^{(2)}$ are components of $\Gamma^{(2)}$ (that is, $\Gamma^{(2)}$ is effectively an ancestral graph of $\Gamma_a^{(1)}$.) This implies that a complete family of ancestors of Γ is $\mathbb{A}(\Gamma) = \{G_1, \Gamma_a^{(2)}, \Gamma_b^{(1)}\}$. If G_1 is exposed, we are done. Suppose G_1 is not exposed. Since G_1 is exposed in $\Gamma_a^{(1)}$ but not in Γ, then by Lemma 3.6 the breeding operation that outputs Γ involves elements of $K(G_1)$ in $\Gamma_a^{(1)}$ or an edge adjacent to $K(G_1)$ whose endpoint is a bridge. By Lemma 3.7, this implies that G_2 is an exposed gene in Γ, and hence at least one of G_1 and G_2 is exposed in Γ. $\square$

Proposition 3.11. *Consider graph* $\Gamma_a^{(1)}$ *with two exposed gene* G_1 *and* G_2 *and no parthenogenic objects. Suppose a parthenogenic operation* $\mathscr{P}$ *is applicable on* $\Gamma_a^{(1)}$ *outputting the graph* $\Gamma_p^{(1)}$ *with a parthenogenic object S. Now consider a breeding operation* $\mathscr{B}$ *involving* $\Gamma_p^{(1)}$ *and another graph* $\Gamma_b^{(1)}$ *such that the resultant graph* Γ *does not contain any parthenogenic objects. Then at least one of* G_1 *or* G_2 *is an exposed gene of* Γ.

Remark 3.26. The rather particular condition described in this proposition arises in the proof of the next proposition, so we handle it in advance here. Roughly speaking, we handle the situation where we start with a graph with no parthenogenic objects and at least two exposed genes. A parthenogenic object is introduced into the graph by a parthenogenic operation, and is then immediately destroyed by a breeding operation. We prove that in this case, the resultant graph also has two exposed genes.

Proof. If both G_1 and G_2 are exposed in $\Gamma_p^{(1)}$ then by Proposition 3.10 at least one of G_1 or G_2 will be exposed in Γ. So it is sufficient to consider the case where at least one of G_1 or G_2 is not exposed in $\Gamma_p^{(1)}$. Without loss of generality, let G_1 be a gene that is not exposed in $\Gamma_p^{(1)}$.

Since G_1 is an exposed gene of $\Gamma_a^{(1)}$, there exists a graph $\Gamma_a^{(2)}$ such that $\mathbb{A}(\Gamma_a^{(1)}) = \{G_1, \Gamma_a^{(2)}\}$ is a complete family of ancestors of $\Gamma_a^{(1)}$. By Corollary 3.1, $\mathbb{A}(\Gamma_a^{(1)})$ is also a complete family of ancestors of $\Gamma_p^{(1)}$. Since Γ is obtained by a breeding operation involving $\Gamma_p^{(1)}$ and $\Gamma_b^{(1)}$, then a complete family of ancestors of Γ is

$$\mathbb{A}(\Gamma) = \mathbb{A}(\Gamma_a^{(1)}) \cup \{\Gamma_b^{(1)}\} = \{G_1, \Gamma_a^{(2)}, \Gamma_b^{(1)}\}.$$

Then by Corollary 3.3, one of the following is true:

   (i) There are two irreducible 3-crackers C_1 and C_2 that separate cores induced by $\mathbb{A}(\Gamma)$ that share an edge e_1. In this case, two of the three ancestors in $\mathbb{A}(\Gamma)$ can be obtained directly by an inverse breeding operation $(\hat{\mathscr{B}}^{-1}, \Gamma, C_1)$ and $\hat{\mathscr{B}}^{-1}(\Gamma, C_2)$ respectively.

  (ii) There is an edge e adjacent to two irreducible 1-crackers, C_1 and C_2, that separate cores induced by $\mathbb{A}(\Gamma)$. In this case, two of the three ancestors in $\mathbb{A}(\Gamma)$ can be obtained directly by an inverse breeding operation, $\hat{\mathscr{B}}^{-1}(\Gamma, C_1)$ and $\hat{\mathscr{B}}^{-1}(\Gamma, C_2)$ respectively.

 (iii) For each of the ancestor in $\mathbb{A}(\Gamma)$, there exists an inverse breeding operation that can isolate the ancestor (as a component of the resultant ancestral graph).

If item (iii) is true, G_1 is an exposed gene of Γ. Note that items (i) and (ii) describe cases 5 and 6 of Proposition 3.3, which are considered in the proof of Proposition 3.4 in Appendix A.2. It is possible to show that if item (i) or (ii) is valid, conditions of this propositions are violated; there is no inverse parthenogenic operation applicable on Γ that create the parthenogenic object S in the resultant graph. In other words, by inspecting the two cases in a similar manner as in Cases 5 and 6 of in the proof of Proposition 3.4 and performing all possible inverse breeding operations, it is straightforward to show that in these two cases there is no inverse breeding operation applicable on Γ such that:

(1) $\Gamma_p^{(1)}$ with a parthenogenic object S is isolated as a component of the resultant ancestral graph,
(2) there exists an inverse parthenogenic operation applicable on $\Gamma_p^{(1)}$ that preserves cores induced by (Γ) and removes S.

Therefore, item (iii) must be true, and so G_1 is an exposed gene of Γ.   □

Now we can prove the following vital proposition, to be used in proving the main result of this chapter.

Proposition 3.12. *Consider a connected cubic graph Γ with $n > 1$ genes and no parthenogenic object. Then Γ has at least two exposed genes.*

Remark 3.27. The primary use of this result will be to imply that it is always possible to identify an ancestor of a descendant graph (without parthenogenic objects) that has one less gene than the descendant. This will enable us to use an inductive argument to prove the uniqueness result. Although this proposition only holds for graphs without parthenogenic objects, by Corollary 3.1 we only need to consider the uniqueness theorem for these graphs.

Proof. First note that the proposition is obvious for $n = 2$, and by Corollary 3.3, it is true for $n = 3$.

Let us assume that the proposition is true for $n < k$, and prove that it is then true for $n = k$. Consider Γ with k genes, Since Γ is a descendant, there exists an inverse breeding operation applicable to Γ which outputs the ancestral graph $\Gamma^{(1)} = \Gamma_p^{(1)} \cup \Gamma_b^{(1)}$ where $\Gamma_p^{(1)}$ and $\Gamma_b^{(1)}$ are components of $\Gamma^{(1)}$. To complete the proof, we first show that there exists a gene (not necessarily exposed) of $\Gamma_p^{(1)}$ that will be exposed in Γ. Then similarly, there exists a gene of $\Gamma_b^{(1)}$ that will be exposed in Γ proving the existence of two exposed genes in Γ.

If $\Gamma_p^{(1)}$ is a gene, then it is an exposed gene of Γ and we are done. Instead, suppose $\Gamma_p^{(1)}$ is not a gene. Then by Corollary 3.2, $\Gamma_p^{(1)}$ contains at most one parthenogenic object. First, suppose $\Gamma_p^{(1)}$ does not contain a parthenogenic object. It is obvious the number of genes for $\Gamma_p^{(1)}$ is less than k. Then, by our assumption, $\Gamma_p^{(1)}$ has two exposed genes G_1 and G_2. By Proposition 3.10, one of the two exposed genes, is an exposed gene of Γ. Instead, suppose $\Gamma_p^{(1)}$ contains one parthenogenic object. Then we can remove this parthenogenic object by an inverse parthenogenic operation and obtain $\Gamma_a^{(1)}$ with no parthenogenic objects. Clearly, $\Gamma_a^{(1)}$ has less than k genes, and so by assumption, $\Gamma_a^{(1)}$ must have at least two exposed genes, say G_1 and G_2. Then by Proposition 3.11, at least one of G_1 and G_2 are exposed in Γ. Therefore, there is always a gene of $\Gamma_p^{(1)}$ that is exposed in Γ.

Similarly, we can show that there exists a gene of $\Gamma_b^{(1)}$ that is exposed in Γ proving the existence of at least two exposed genes in Γ.   □

3.2.5 Uniqueness of the Complete Family of Ancestor Genes

We are now ready to complete the proof of the uniqueness of the complete family of ancestor genes. Consider a descendent graph Γ and an exposed gene G of Γ. We will first show that if a sequence of inverse operations is performed on Γ, outputting the ancestral graph $\Gamma^{(\cdot)}$, then G is either a component of $\Gamma^{(\cdot)}$ or an exposed gene of $\Gamma^{(\cdot)}$. This result will allow us to use induction on the number of genes to prove the uniqueness of the complete family of ancestor genes.

Proposition 3.13. *Consider a graph Γ with an exposed gene G. Let C_1 be the irreducible cubic cracker such that $\hat{\mathscr{B}}^{-1}(\Gamma, C_1)$ isolates G as a component of the resultant ancestral graph. Consider another irreducible cubic cracker $C_2 \neq C_1$ and let the ancestral graph $\Gamma^{(1)} = \hat{\mathscr{B}}^{-1}(\Gamma, C_2)$. Then G is an exposed gene of $\Gamma^{(1)}$ as well.*

Remark 3.28. If a gene is exposed, it is clear that there is a unique inverse breeding operation that will isolate it. This result ensures that if any other inverse breeding operation is performed, that gene remains exposed.

Proof. Suppose the deletion of C_1 from Γ separates two components Γ_1 and Γ_2. Our aim is to invoke Proposition 3.5 for crackers C_1 and C_2.

Therefore we first need to show that one one subgraph separated by the deletion of C_1 does not contain any edges of C_2, and vice versa.

Let us choose any complete family of ancestor genes containing G (it is clear from the definition of exposed genes that such a complete family of ancestor genes exists). Then without loss of generality, assume subgraph Γ_1 corresponds to G in the sense that Γ_1 contains $K(G)$, the core induced by G. There are now two possible cases. First, if C_1 is a 2-cracker or a 3-cracker, then it is clear that deleting this cracker isolates $K(G)$, and so by definition of inverse type 2 and inverse type 3 breeding, Γ_1 is precisely $K(G)$ in this case. Second, if C_1 is a 1-cracker, then deleting this cracker isolates the subgraph Γ_1, which by definition of inverse type 1 breeding is made up of the union of two edges adjacent to the 1-cracker, and $K(G)$. We consider these two situations separately.

(i) If C_1 is a 2-cracker or a 3-cracker, then $\Gamma_1 = K(G)$, and by Corollary 3.4 we can immediately conclude that Γ_1 does not contain any cracker edge. Therefore, all edges (or remaining edges) of C_2, after deleting C_1 from Γ, must lie in Γ_2.

(ii) If C_1 is a 1-cracker, then Γ_1 differs from the core induced by G by the presence of edges $e_1 = (u, v)$ and $e_2 = (u, w)$ that were originally adjacent to C_1. By Corollary 3.4, $K(G)$ does not contain any cubic cracker edge, so we only need to check if e_1 or e_2 are cubic cracker edges. Suppose e_1 is a cubic cracker edge. Then consider the endpoints u and v of e_1. There exists a path starting from u, passing through e_2 (which cannot be in a cubic cracker with e_1 since the two edges are adjacent) and then some edges $K(G)$ (none of which are cubic cracker edges) before ending in v. Therefore the cracker containing e_1 does not separate vertices v and u, implying that e_1 is an unnecessary member of the cracker. Since crackers are minimal cutsets, this is a contradiction, and so we conclude that e_1 is not a cubic cracker edge. Using an identical argument, we also conclude that e_2 is not a cracker edge. Therefore, there are no cubic cracker edges in Γ_1. Then all edges (or remaining edges) of C_2, after deleting C_1 from Γ, are in Γ_2.

Therefore it becomes clear that one subgraph separated by the deletion of C_1 does not contain any edges of C_2, and vice versa. Then by Proposition 3.5, if we perform the inverse operation $\hat{\mathscr{B}}^{-1}(\Gamma, C_2)$, resulting in $\Gamma^{(1)}$, then it is still possible to perform an inverse operation on $\Gamma^{(1)}$ that outputs G, and so G is still exposed in $\Gamma^{(1)}$. $\square$

Proposition 3.14. *Consider Γ with $n > 1$ genes. Suppose $\mathbb{A}(\Gamma) = \{G, \Gamma_a\}$ is a complete family of ancestors of Γ such that G is a gene. Suppose an inverse breeding operation is performed on Γ resulting in the ancestral graph $\Gamma^{(1)}$. Then one of the following must hold:*

(i) $\Gamma^{(1)} = \Gamma_a \cup G$, or
(ii) there exist graphs Γ_2, Γ_3 and Γ_b such that $\Gamma^{(1)} = \Gamma_b \cup \Gamma_3$, $\mathbb{A}(\Gamma_a) = \{\Gamma_2, \Gamma_3\}$ is a complete family of ancestors of Γ_a, and $\mathbb{A}(\Gamma_b) = \{\Gamma_2, G\}$ is a complete family of ancestors of Γ_b.

Remark 3.29. This result implies that, if we highlight a particular ancestor gene of a graph, that any inverse operation which does not immediately isolate this gene instead outputs ancestors which are related to that gene in a very particular way. Since this result holds for any exposed gene in the original graph, it is quite powerful and will be used in the proof of the main theorem.

Proof. Since $\mathbb{A}(\Gamma)$ is a complete family of ancestor of Γ, it is clear that there exists a breeding operation which may be performed on G and Γ_a which results in Γ. It is also clear that Γ_a has $n - 1$ genes.

If an inverse breeding operation on Γ results in G and Γ_a then item (i) is valid. Consider the alternative situation. Clearly in this case $n > 2$ (or else there would not have been a second inverse operation applicable), and it is similarly obvious that if G is not obtained, then neither is Γ_a. In that case, $\Gamma^{(1)}$ is made up of two graphs, both of which are different from G and Γ. Consistent with the notation given in the statement of the proposition, we call these two graphs Γ_b and Γ_3. Since $n > 3$ we know that it is impossible for both Γ_b and Γ_3 to be genes, so without loss of generality we assume that Γ_b is a descendant, whereas Γ_3 could be a gene or a descendant. Then by Proposition 3.13, G is an exposed gene in $\Gamma^{(1)}$. Without loss of generality, we assume that G lies in Γ_b. Then there exists an inverse breeding operation that can be performed on $\Gamma^{(1)}$, involving a cubic cracker of Γ_b, resulting in $\Gamma^{(2)}$ such that one of the resultant components is G. Another component of $\Gamma^{(2)}$ is obviously Γ_3 and we denote the third component by Γ_2. Note that $\{G, \Gamma_2\}$ is a complete family of ancestors of Γ_b. We now need to show that $\{\Gamma_2, \Gamma_3\}$ is a complete family of ancestors of Γ_a. By Proposition 3.4 we can separate the cores induced by $\{G, \Gamma_2, \Gamma_3\}$ in any order and the result is unchanged. Therefore we can first separate G which results in

$\Gamma_a \cup G$. The second inverse breeding operation would separate the cores induced by Γ_2 and Γ_3 resulting in G, Γ_2 and Γ_3. Therefore $\{\Gamma_2, \Gamma_3\}$ a complete family of ancestors of Γ_a and part (ii) of the proposition holds. □

Lemma 3.8. *Consider Γ with a unique complete family of ancestor genes $\mathbb{G}(\Gamma)$. Consider any complete family of ancestors of Γ, containing Γ_1, Γ_2, ..., Γ_n, each themselves having a unique complete family of ancestor genes $\mathbb{G}(\Gamma_1)$, $\mathbb{G}(\Gamma_2)$, ..., $\mathbb{G}(\Gamma_n)$ respectively. Then*

$$\mathbb{G}(\Gamma) = \mathbb{G}(\Gamma_1) \cup \mathbb{G}(\Gamma_2) \cup \ldots \cup \mathbb{G}(\Gamma_n).$$

Remark 3.30. Even if a graph has a unique complete family of ancestor genes, it may have many different complete families of ancestors. This result ensures that for any of those families, if the ancestors also have unique complete families of ancestor genes, then the we can merely find their ancestor genes and take the union. This result will be used to finish the proof of the main theorem.

Proof. Consider gene $G \in \mathbb{G}(\Gamma)$. Since G should be obtained after any sequence of inverse operations, of sufficient length, on Γ, it must be possible to obtain G after a sequence of inverse operations on one of $\Gamma_1, \Gamma_2, \ldots, \Gamma_n$. Therefore

$$\mathbb{G}(\Gamma) \subset \mathbb{G}(\Gamma_1) \cup \mathbb{G}(\Gamma_2) \cup \ldots \cup \mathbb{G}(\Gamma_n).$$

Furthermore, any gene obtained after a sequence of inverse operations on $\Gamma^{(\cdot)} = \Gamma_1 \cup \Gamma_2 \cup \ldots \cup \Gamma_n$ is a gene of Γ. Therefore

$$\mathbb{G}(\Gamma_1) \cup \mathbb{G}(\Gamma_2) \cup \ldots \cup \mathbb{G}(\Gamma_n) \subset \mathbb{G}(\Gamma),$$

thereby completing the proof. □

Suppose $\mathbb{A}(\Gamma) = \{\Gamma_1, \Gamma_2\}$ is a complete family of ancestors of Γ where Γ_1 and Γ_2 both have unique families of ancestor genes. Then by the nature of inverse operations any sequence of inverse operations on the ancestral graph $\Gamma^{(1)} = \Gamma_1 \cup \Gamma_2$ results in a unique complete family of ancestors. Keeping this in mind, we move on to proving the main theoretical result of this study, the uniqueness of the complete family of ancestor genes.

Theorem 3.3. *First, any cubic graph has a complete family of ancestor genes and second, a complete family of ancestor genes for a given connected cubic graph Γ is unique.*

Remark 3.31. Finally, we are in a position to prove the main result. An inductive argument will be used whereby we assume that the result is true for all graphs without parthenogenic objects and containing $n-1$ genes, and use that to prove the result for graphs without parthenogenic objects and containing n genes.

Proof. The first part of the theorem follows from Theorem 1.1. To show the second part, we need only consider graphs with no parthenogenic objects, because by

Corollary 3.1 if any are present we can always initially perform a sequence of inverse parthenogenic operations to remove them, and the resultant graph contains all the ancestor genes of the original graph.

Suppose Γ has n genes and no parthenogenic objects. If $n = 2$ the theorem is trivial, and by Corollary 3.6 when $n = 3$ the theorem is also valid. We will use induction to show that it holds for any positive natural number n. Suppose all cubic graphs with $n - 1$ genes have unique complete families of ancestor genes. Then we need to show that any cubic graph with n genes also has a unique complete family of ancestor genes. By Proposition 3.12, any such graph Γ has at least two exposed genes, and so in particular there exists a gene G and a graph Γ_a with $n - 1$ genes such that a complete family of ancestors $\mathbb{A}(\Gamma) = (G, \Gamma_a)$ can be obtained by a particular inverse breeding operation on Γ. Then we consider the two possible cases that may arise when we perform any inverse breeding operation on Γ.

Case 1: An inverse breeding operation is performed, resulting in $G \cup \Gamma_a$. Proposition 3.12 implies that Γ_a has $n - 1$ ancestor genes, and therefore by the inductive assumption, Γ_a has a unique complete family of ancestor genes. Therefore the only complete family of ancestor genes that could be eventually obtained by continuing on from this case is $\mathbb{G}(\Gamma_a) \cup G$.

Case 2: An inverse breeding operation is performed, resulting in $\Gamma^{(1)} \neq G \cup \Gamma_a$. Suppose $\Gamma^{(1)} = \Gamma_b \cup \Gamma_3$. By Proposition 3.14, there exists Γ_2 such that $\mathbb{A}(\Gamma_a) = \{\Gamma_2, \Gamma_3\}$ is a complete family of ancestors of Γ_a, and $\mathbb{A}(\Gamma_b) = \{\Gamma_2, G\}$ is a complete family of ancestors of Γ_b. By the inductive assumption, Γ_b and Γ_3 each have a unique complete family of ancestor genes $\mathbb{G}(\Gamma_b)$ and $\mathbb{G}(\Gamma_3)$ respectively. Since $\Gamma^{(1)}$ contains two connected components, it is clear that any inverse operation performed on $\Gamma^{(1)}$ only interacts with one of Γ_b and Γ_3 at a time, and the ancestor genes obtained from each of these is not influenced by the inverse operations performed on the other. Then, since Γ_b and Γ_3 have unique complete families of ancestor genes, it is clear that the unique complete family of ancestor genes that may be obtained from this stage is $\mathbb{G}(\Gamma^{(1)}) = \mathbb{G}(\Gamma_b) \cup \mathbb{G}(\Gamma_3)$.

Then, from Lemma 3.8, we also have

$$\mathbb{G}(\Gamma_b) = \mathbb{G}(\Gamma_2) \cup G,$$

where $\mathbb{G}(\Gamma_2)$ is the complete family of ancestor genes of Γ_2. Therefore, the genes obtained in this case are

$$\mathbb{G}(\Gamma^{(1)}) = \mathbb{G}(\Gamma_3) \cup \mathbb{G}(\Gamma_2) \cup G = \mathbb{G}(\Gamma_a) \cup G.$$

Since Γ_a and G are fixed graphs, the complete family of ancestor genes obtained by this case is unique.

Cases 1 and 2 both result in the same complete family of ancestor genes, therefore the complete family of ancestor genes for a given connected cubic graph Γ is unique.
□

With the proof of the main theorem complete, we conclude this manuscript with some final reflections on the value of the result. In Chap. 1 we proved that every cubic graph can be viewed as either being a gene, or a graph which has been constructed from a complete family of ancestor genes, and that one such complete family may be identified in polynomial time. Theorem 3.3 implies that the family of genes used to construct the graph is unique, so no concern need be had as to the order of operations by which such a complete family is found. It is also our belief that some of the results in this chapter imply that algorithms that identify ancestor genes can, in large part, be parallelised. Since every descendant can now be viewed as being constructed from a unique complete family of ancestor genes, the results of Chap. 2 become more potent. Analysis of a descendant can be reduced to analysis of a family of smaller cubic graphs which is unique.

It is our hope that other inherited properties of descendants will be discovered. Certainly it is our belief that one-directional inheritance properties abound, and more two-directional properties are likely to exist as well. The results presented in this manuscript provide a framework for constructing graphs with desirable properties, whether they be for a practical application, or in the search for a counterexample to a conjecture. They also provide a framework for verifying properties, or explaining the similarities or differences within a set of cubic graphs.

Appendix A
Completed Proofs from Chapter 3

A.1 Full Proof of Theorem 3.2

We restate Theorem 3.2 here.

Theorem A.1. *Consider a cubic graph Γ containing a parthenogenic object S. Let Γ_p be the graph obtained after performing the inverse parthenogenic operation:*

$$\Gamma_p = \hat{\mathscr{B}}^{-1}(\Gamma, S).$$

If $\Gamma^{(\cdot)}$ is an ancestral graph of Γ which is neither Γ_p, nor an ancestral graph of Γ_p, then $\Gamma^{(\cdot)}$ contains at least one parthenogenic object, and there exists a sequence of inverse parthenogenic operations that, when performed on $\Gamma^{(\cdot)}$, produces an ancestral graph of Γ_p.

Proof. By Proposition 1.4, there exists an applicable inverse operation on the descendant graph Γ. Suppose $\Gamma^{(\cdot)}$ is obtained after a sequence of n inverse operations on Γ. We will use induction on n to show the validity of the theorem.

Base Step:

We first consider the case where $n = 1$. In other words, we consider the situation where only a single inverse operation is performed on Γ to obtain $\Gamma^{(1)}$. We show that in all cases there exists $\Gamma^{(2)}$, obtained by a sequence of (no more than two) inverse parthenogenic operations on $\Gamma^{(1)}$, that is an ancestral graph of Γ_p.

Case 1: Suppose $\Gamma^{(1)} = \hat{\mathscr{B}}^{-1}(\Gamma, C)$ where $C \in \Gamma$ is an irreducible cubic cracker, and C does not share any edges with S, or with those edges adjacent to S. Therefore, if the appropriate inverse parthenogenic operation is performed on Γ to produce Γ_p, cracker C remains unaltered, and therefore irreducible in Γ_p. Likewise, if $\hat{\mathscr{B}}^{-1}(\Gamma, C)$ is performed, the parthenogenic object S remains in $\Gamma^{(1)}$. Then it is clear that the order of the two inverse operations is unimportant. Such a case is illustrated in Fig. A.1.

© Springer International Publishing Switzerland 2016
P. Baniasadi et al., *Genetic Theory for Cubic Graphs*, SpringerBriefs
in Operations Research, DOI 10.1007/978-3-319-19680-0

Therefore, an inverse breeding operation on C is possible in Γ_p resulting in an ancestral graph, say $\Gamma_p^{(2)}$, and an inverse parthenogenic operation on S is possible in $\Gamma^{(1)}$ which results in the same ancestral graph.

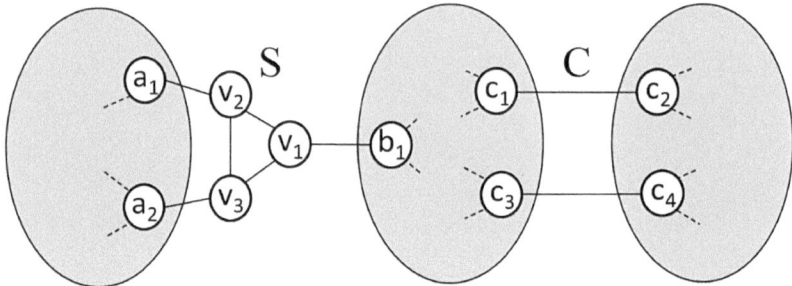

Fig. A.1 A graph containing irreducible cubic cracker C and parthenogenic object S described in Case 1

Case 2: Suppose $\Gamma^{(1)} = \hat{\mathscr{B}}^{-1}(\Gamma, S_1)$ where $S_1 \neq S$ is a parthenogenic object in Γ, and its adjacent edges are distinct from the adjacent edges of S. Such a case is illustrated in Fig. A.2. Similarly to Case 1, the order of inverse parthenogenic operations is unimportant, and therefore using analogous arguments, we conclude that the graph $\Gamma^{(2)}$ obtained by $\hat{\mathscr{B}}^{-1}(\Gamma^{(1)}, S)$ is an ancestor of Γ_p.

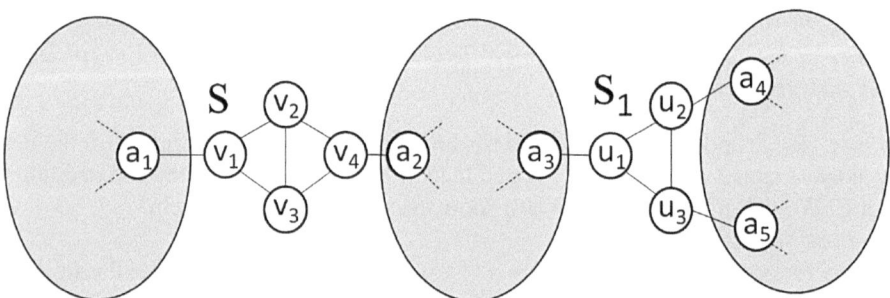

Fig. A.2 A graph containing the two parthenogenic objects S and S_1 described in Case 2

Case 3: Suppose $S_1 \neq S$ is a parthenogenic object where $\Gamma^{(1)} = \hat{\mathscr{B}}^{-1}(\Gamma, S_1)$, and there exists at least one adjacent edge e of S which is also an adjacent edge of S_1. By inspection, all of the possible cases are the following:

(i) Parthenogenic objects S and S_1 are both parthenogenic diamonds and are both adjacent to a reducible 1-cracker e.

(ii) Parthenogenic objects S and S_1 are both parthenogenic triangles and are both adjacent to a reducible 1-cracker e.

(iii) Parthenogenic objects S and S_1 are a parthenogenic diamond and a parthe-
     nogenic triangle (in either order), both adjacent to a reducible 1-cracker e.
(iv) Parthenogenic objects S and S_1 are parthenogenic both parthenogenic bridges
     and are both adjacent to a reducible 2-cracker C, containing e.

The above four cases are illustrated in Fig. A.3.

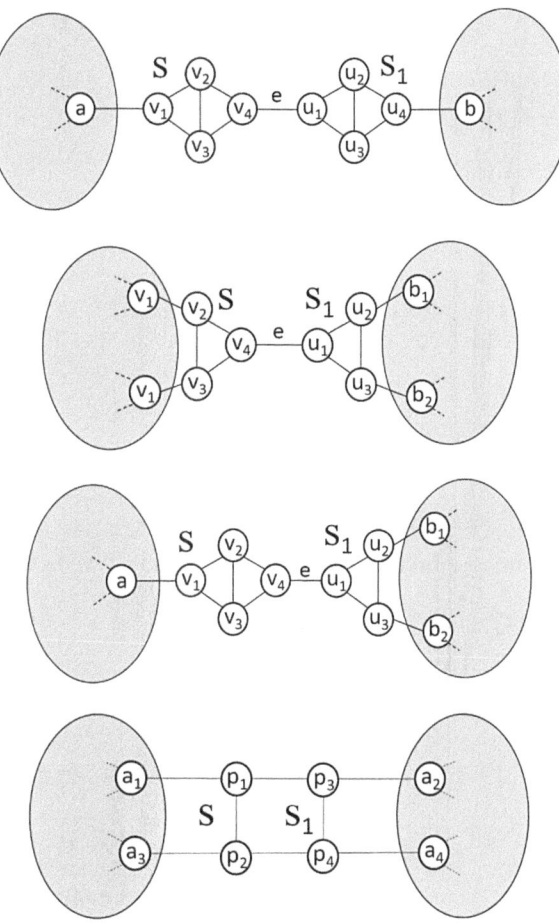

Fig. A.3 The four situations illustrate Γ where it contains two parthenogenic objects that share
adjacent edges

It can be checked that in all of the above cases

$$\hat{\mathscr{B}}^{-1}(\Gamma^{(1)}, S) = \hat{\mathscr{B}}^{-1}(\Gamma_p, S_1) := \Gamma_p^{(2)}.$$

Case 4: Suppose $\Gamma^{(1)} = \hat{\mathscr{B}}^{-1}(\Gamma, C)$ where C is an irreducible cubic cracker
that shares edges with those adjacent to the parthenogenic object S. Note that by
Lemma 3.1, exactly one edge of C must be adjacent to S. Therefore, C is not a

bridge as it is an irreducible cracker. From Proposition 3.1, if we remove S, its adjacent edges, and the edges of C, there will be three connected components Γ_1, Γ_2 and Γ_3 that are subgraphs of Γ. Note that it is not possible for C to share an edge adjacent to a parthenogenic diamond, as both of these edges are 1-crackers, and C must be a minimal cutset. Therefore S is either a parthenogenic bridge or a parthenogenic triangle.

Parthenogenic object S is a parthenogenic bridge:

Let S be a parthenogenic bridge containing edge (p_1, p_2). Also let the four adjacent edges to S be (v_1, p_1), (v_2, p_2), (u_1, p_1) and (s_1, p_2). The irreducible cubic cracker C could be a 2-cracker or a 3-cracker (see Figs. A.4 and A.5 respectively).

Case 4.1: If C is a 2-cracker then without loss of generality,

$$V(\Gamma) = V(\Gamma_1) \cup V(\Gamma_2) \cup V(\Gamma_3) \cup \{p_1, p_2\},$$
$$E(\Gamma) = E(\Gamma_1) \cup E(\Gamma_2) \cup E(\Gamma_3)$$
$$\cup \{(v_1, p_1), (v_2, p_2), (u_1, p_1), (s_1, p_2), (s_2, u_2), (p_1, p_2)\},$$

where $v_1, v_2 \in V(\Gamma_1)$, $u_1, u_2 \in V(\Gamma_2)$, $s_1, s_2 \in V(\Gamma_3)$, and $C = \{(s_1, p_2), (s_2, u_2)\}$ is the irreducible 2-cracker. This case is illustrated in Fig. A.4. Then $\Gamma_p = \hat{\mathscr{B}}^{-1}(\Gamma, S)$ is a graph such that

$$V(\Gamma_p) = V(\Gamma_1) \cup V(\Gamma_2) \cup V(\Gamma_3),$$
$$E(\Gamma_p) = E(\Gamma_1) \cup E(\Gamma_2) \cup E(\Gamma_3) \cup \{(v_1, u_1), (v_2, s_1), (s_2, u_2)\}.$$

Obviously, $C_p = \{(v_2, s_1), (u_2, s_2)\}$ is an irreducible 2-cracker in Γ_p.

Performing the inverse breeding operation $\hat{\mathscr{B}}^{-1}(\Gamma, C)$ we obtain $\Gamma^{(1)}$ where

$$V(\Gamma^{(1)}) = V(\Gamma_1) \cup V(\Gamma_2) \cup V(\Gamma_3) \cup \{p_1, p_2\}$$
$$E(\Gamma^{(1)}) = E(\Gamma_1) \cup E(\Gamma_2)$$
$$\cup E(\Gamma_3) \cup \{(v_1, p_1), (v_2, p_2), (u_1, p_1), (u_2, p_2), (p_1, p_2), (s_1, s_2)\},$$

and $S^{(1)} = \{(p_1, p_2)\}$ is a parthenogenic bridge in $\Gamma^{(1)}$. It can be checked that

$$\hat{\mathscr{B}}^{-1}(\Gamma^{(1)}, S^{(1)}) = \hat{\mathscr{B}}^{-1}(\Gamma_p, C_p) := \Gamma_p^{(2)}.$$

Case 4.2: If C is a 3-cracker then without loss of generality,

$$V(\Gamma) = V(\Gamma_1) \cup V(\Gamma_2) \cup V(\Gamma_3) \cup \{p_1, p_2\},$$
$$E(\Gamma) = E(\Gamma_1) \cup E(\Gamma_2) \cup E(\Gamma_3)$$
$$\cup \{(v_1, p_1), (v_2, p_2), (u_1, p_1), (s_1, p_2), (s_2, u_2), (s_3, u_3), (p_1, p_2)\},$$

where $v_1, v_2 \in V(\Gamma_1)$, $u_1, u_2, u_3 \in V(\Gamma_2)$, $s_1, s_2, s_3 \in V(\Gamma_3)$, and $C = \{(s_1, p_2), (s_2, u_2), (s_3, u_3)\}$ is the irreducible 3-cracker. This case is illustrated in Fig. A.5. Then Γ_p would be a graph such that

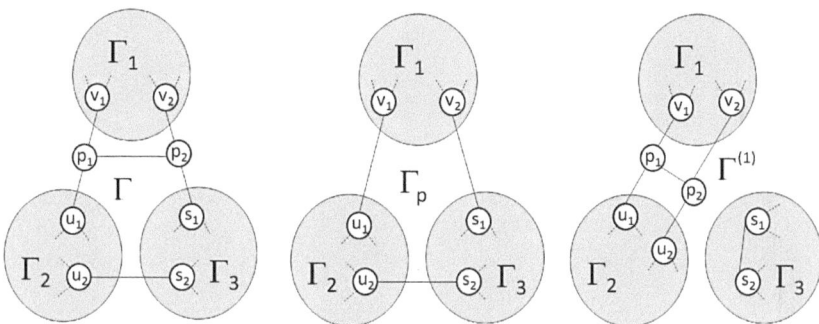

Fig. A.4 The graphs Γ, Γ_p and $\Gamma^{(1)}$ as described in Case 4.1

$$V(\Gamma_p) = V(\Gamma_1) \cup V(\Gamma_2) \cup V(\Gamma_3),$$
$$E(\Gamma_p) = E(\Gamma_1) \cup E(\Gamma_2) \cup E(\Gamma_3) \cup \{(v_1, u_1), (v_2, s_1), (s_2, u_2), (s_3, u_3)\}.$$

Obviously, $C_p = \{(v_2, s_1), (s_2, u_2), (s_3, u_3)\}$ is an irreducible 3-cracker in Γ_p.

Performing the inverse breeding operation $\hat{\mathscr{B}}^{-1}(\Gamma, C)$ we obtain $\Gamma^{(1)}$ where

$$V(\Gamma^{(1)}) = V(\Gamma_1) \cup V(\Gamma_2) \cup V(\Gamma_3) \cup \{p_1, p_2, w_1, w_2\}$$
$$E(\Gamma^{(1)}) = E(\Gamma_1) \cup E(\Gamma_2) \cup E(\Gamma_3) \cup \{(v_1, p_1), (v_2, p_2),$$
$$\cup (u_1, p_1), (u_2, w_1), (u_3, w_1), (w_1, p_2), (s_1, w_2), (s_2, w_2), (s_3, w_2), (p_1, p_2)\},$$

where w_1 and w_2 are new vertices and $S^{(1)} = \{(p_1, p_2)\}$ is a parthenogenic bridge in $\Gamma^{(1)}$. It can be checked that

$$\hat{\mathscr{B}}^{-1}(\Gamma^{(1)}, S^{(1)}) = \hat{\mathscr{B}}^{-1}(\Gamma_p, C_p) := \Gamma_p^{(2)},$$

recalling that when performing inverse breeding operation on the 3-cracker C_p two new vertices, w_1 and w_2 are created.

Parthenogenic object S is a parthenogenic triangle:

Let vertices p_1, p_2 and p_3 be the vertices of the parthenogenic triangle S. Also let the three adjacent edges to S be (v_1, p_1), (u_1, p_2) and (s_1, p_3). The irreducible cubic cracker C could be a 2-cracker, Case 4.3, or a 3-cracker, Case 4.4 (see Figs. A.6 and A.7 respectively).

Case 4.3: If C is a 2-cracker then without loss of generality,

$$V(\Gamma) = V(\Gamma_1) \cup V(\Gamma_2) \cup V(\Gamma_3) \cup \{p_1, p_2, p_3\},$$
$$E(\Gamma) = E(\Gamma_1) \cup E(\Gamma_2) \cup E(\Gamma_3)$$
$$\cup \{(v_1, p_1), (u_1, p_2), (s_1, p_3), (p_1, p_2), (p_2, p_3), (p_3, p_1), (u_2, s_2)\},$$

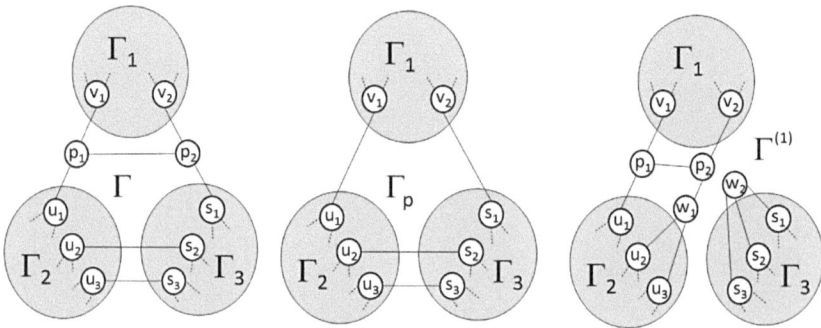

Fig. A.5 The graphs Γ, Γ_p and $\Gamma^{(1)}$ as described in Case 4.2

where $v_1 \in V(\Gamma_1)$, $u_1, u_2 \in V(\Gamma_2)$, $s_1, s_2 \in V(\Gamma_3)$, and $C = \{(u_1, p_2), (u_2, s_2)\}$ is the irreducible 2-cracker. This case is illustrated in Fig. A.6. Then $\Gamma_p = \hat{\mathscr{B}}^{-1}(\Gamma, S)$ is a graph such that

$$V(\Gamma_p) = V(\Gamma_1) \cup V(\Gamma_2) \cup V(\Gamma_3) \cup \{v2\},$$
$$E(\Gamma_p) = E(\Gamma_1) \cup E(\Gamma_2) \cup E(\Gamma_3) \cup \{(v_1, v_2), (u_1, v_2), (s_1, v_2), (u_2, s_2)\}.$$

Obviously, $C_p = \{(u_1, v_2), (u_2, s_2)\}$ is an irreducible 2-cracker in Γ_p.

Performing the inverse breeding operation $\hat{\mathscr{B}}^{-1}(\Gamma, C)$ we obtain $\Gamma^{(1)}$ where

$$V(\Gamma^{(1)}) = V(\Gamma_1) \cup V(\Gamma_2) \cup V(\Gamma_3) \cup \{p_1, p_2, p_3\}$$
$$E(\Gamma^{(1)}) = E(\Gamma_1) \cup E(\Gamma_2) \cup E(\Gamma_3)$$
$$\cup \{(v_1, p_1), (p_1, p_2), (p_2, p_3), (p_3, p_1), (p_2, s_2), (p_3, s_1), (u_1, u_2)\},$$

and $S^{(1)} = \{(p_1, p_2, p_3)\}$ is a parthenogenic triangle in $\Gamma^{(1)}$. It can be checked that

$$\hat{\mathscr{B}}^{-1}(\Gamma^{(1)}, S^{(1)}) = \hat{\mathscr{B}}^{-1}(\Gamma_p, C_p) := \Gamma_p^{(2)}.$$

Case 4.4: If C is a 3-cracker then without loss of generality,

$$V(\Gamma) = V(\Gamma_1) \cup V(\Gamma_2) \cup V(\Gamma_3) \cup \{p_1, p_2, p_3\},$$
$$E(\Gamma) = E(\Gamma_1) \cup E(\Gamma_2) \cup E(\Gamma_3)$$
$$\cup \{(v_1, p_1), (u_1, p_2), (s_1, p_3), (p_1, p_2), (p_2, p_3), (p_3, p_1), (u_2, s_2), (u_3, s_3)\},$$

where $v_1 \in V(\Gamma_1)$, $u_1, u_2, u_3 \in V(\Gamma_2)$, $s_1, s_2, s_3 \in V(\Gamma_3)$, and $C = \{(u_1, p_2), (u_2, s_2), (u_3, s_3)\}$ is the irreducible 3-cracker. This case is illustrated in Fig. A.6. Then $\Gamma_p = \hat{\mathscr{B}}^{-1}(\Gamma, S)$ is a graph such that

$$V(\Gamma_p) = V(\Gamma_1) \cup V(\Gamma_2) \cup V(\Gamma_3) \cup \{v2\},$$
$$E(\Gamma_p) = E(\Gamma_1) \cup E(\Gamma_2) \cup E(\Gamma_3) \cup \{(v_1, v_2), (u_1, v_2), (s_1, v_2), (u_2, s_2), (u_3, s_3)\}.$$

Obviously, $C_p = \{(u_1, v_2), (u_2, s_2), (u_3, s_3)\}$ is an irreducible 3-cracker in Γ_p.

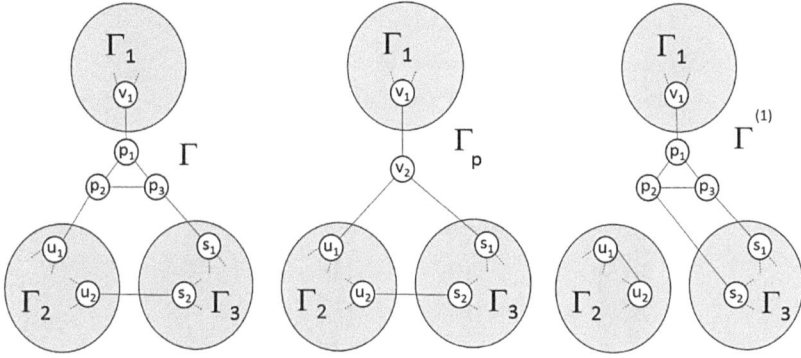

Fig. A.6 The graphs Γ, Γ_p and $\Gamma^{(1)}$ as described in Case 4.3

Performing the inverse breeding operation $\hat{\mathscr{B}}^{-1}(\Gamma, C)$ we obtain $\Gamma^{(1)}$ where

$$V(\Gamma^{(1)}) = V(\Gamma_1) \cup V(\Gamma_2) \cup V(\Gamma_3) \cup \{p_1, p_2, p_3, w_1, w_2\}$$
$$E(\Gamma^{(1)}) = E(\Gamma_1) \cup E(\Gamma_2) \cup E(\Gamma_3)$$
$$\cup \{(v_1, p_1), (p_1, p_2), (p_2, p_3), (p_3, p_1), (p_2, w_2), (p_3, s_1), (w_2, s_2), (w_2, s_3)\},$$

and $S^{(1)} = \{(p_1, p_2, p_3)\}$ is a parthenogenic triangle in $\Gamma^{(1)}$. It can be checked that

$$\hat{\mathscr{B}}^{-1}(\Gamma^{(1)}, S^{(1)}) = \hat{\mathscr{B}}^{-1}(\Gamma_p, C_p) := \Gamma_p^{(2)}.$$

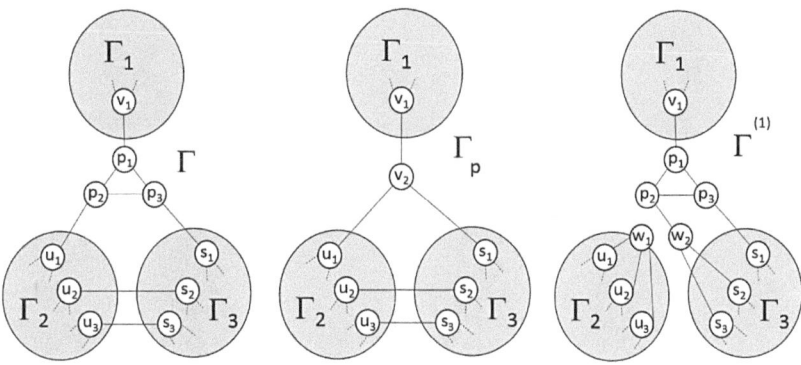

Fig. A.7 The graphs Γ, Γ_p and $\Gamma^{(1)}$ as described in Case 4.4

Case 5: An edge of S is shared with C. This case cannot occur for a parthenogenic triangle or a parthenogenic diamond. Therefore S is a parthenogenic bridge. By Lemma 3.2, C is a 3-cracker. From the proof of Lemma 3.2, Γ can be described as

$$V(\Gamma) = V(\Gamma_1) \cup V(\Gamma_2) \cup V(\Gamma_3) \cup V(\Gamma_4) \cup \{p_1, p_2\},$$
$$E(\Gamma) = E(\Gamma_1) \cup E(\Gamma_2) \cup E(\Gamma_3) \cup E(\Gamma_4)$$
$$\cup \{(u_1, p_1), (p_1, v_1), (u_2, u_3), (p_1, p_2), (v_2, v_3), (u_4, p_2), (p_2, v_4)\},$$

where $u_1, u_2 \in V(\Gamma_1)$, $u_3, u_4 \in V(\Gamma_2)$, $v_1, v_2 \in V(\Gamma_3)$, $v_3, v_4 \in V(\Gamma_4)$, $S = \{(p_1, p_2)\}$ and $C = \{(u_2, u_3), (p_1, p_2), (v_2, v_3)\}$ is the 3-cracker. This case is illustrated in Fig. A.8. Then Γ_p would be a graph such that

$$V(\Gamma) = V(\Gamma_1) \cup V(\Gamma_2) \cup V(\Gamma_3) \cup V(\Gamma_4)$$
$$E(\Gamma) = E(\Gamma_1) \cup E(\Gamma_2) \cup E(\Gamma_3) \cup E(\Gamma_4)$$
$$\cup \{(u_1, v_1), (u_2, u_3), (v_2, v_3), (u_4, v_4)\},$$

Obviously, $C_p = \{(u_2, u_3), (v_2, v_3)\}$ is an irreducible 2-cracker in Γ_p.

Performing the inverse breeding operation $\mathscr{B}^{-1}(\Gamma, C)$ we obtain $\Gamma^{(1)}$ where

$$V(\Gamma^{(1)}) = V(\Gamma_1) \cup V(\Gamma_2) \cup V(\Gamma_3) \cup \{p_1, p_2, p_3, p_4\}$$
$$E(\Gamma^{(1)}) = E(\Gamma_1) \cup E(\Gamma_2) \cup E(\Gamma_3) \cup \{(u_1, p_1), (v_1, p_1),$$
$$\cup (p_1, p_3), (u_2, p_3), (p_3, v_2), (u_3, p_4), (p_4, v_3), (p_4, p_2), (u_4, p_2), (p_2, v_4)\},$$

where p_3 and p_4 are new vertices. It is straightforward to verify that $S^{(1)} = \{(p_1, p_3)\}$ and $S^{(2)} = \{(p_2, p_4)\}$ are parthenogenic bridges in $\Gamma^{(1)}$. Then, it can be checked that

$$\mathscr{B}^{-1}(\mathscr{B}^{-1}(\Gamma^{(1)}, S^{(1)}), S^{(2)}) = \mathscr{B}^{-1}(\Gamma_p, C_p) := \Gamma_p^{(2)}.$$

Note that in this case, we required a sequence of two inverse parthenogenic operations to obtain an ancestral graph of Γ_p.

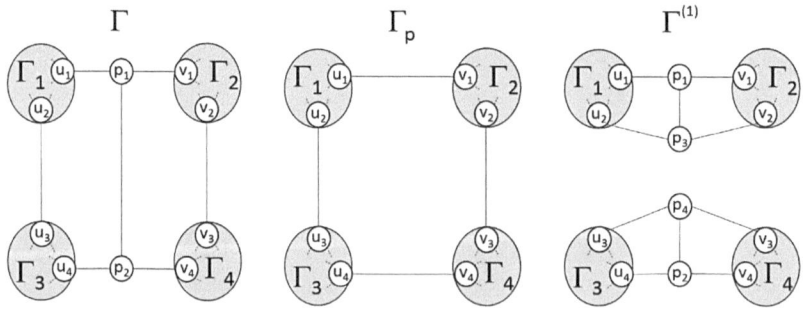

Fig. A.8 The graphs Γ, Γ_p and $\Gamma^{(1)}$ as described in Case 5

Inductive step:

For cases 1–4, the graph $\Gamma_p^{(2)}$ is obtained by performing an inverse parthenogenic operation on $\Gamma^{(1)}$. Also, for case 5 it is easy to confirm that $\Gamma_p^{(2)}$ can be obtained after a sequence of two inverse parthenogenic operations is performed on $\Gamma^{(1)}$.

Now, assume the theorem is true for a particular value $n = k$. That is, for any ancestral graph $\Gamma^{(k)}$ of Γ obtained after k inverse operations which is not Γ_p or an ancestral graph of Γ_p, there exists a sequence of inverse parthenogenic operations which may be performed on $\Gamma^{(k)}$ that results in an ancestral graph of Γ_p. Then suppose some inverse operation is performed on $\Gamma^{(k)}$, producing ancestral graph $\Gamma^{(k+1)}$ which is not an ancestral graph of Γ_p. Then this situation is identical to the base case, except instead of starting with Γ, $\Gamma^{(1)}$ and Γ_p, we are now starting with $\Gamma^{(k)}$, $\Gamma^{(k+1)}$, and either an ancestral graph of Γ_p or a graph from which an ancestral graph of Γ_p may be obtained by performing a sequence of inverse parthenogenic operations. Call this latter graph Γ_p^*. Through the same arguments as those used for the base case, it is then possible to perform one or two inverse parthenogenic operations on $\Gamma^{(k+1)}$ to obtain Γ_p^*, and then proceed to perform a sequence of inverse parthenogenic operations on Γ_p^* to obtain an ancestral graph of Γ_p. Therefore if the result is true for $n = k$, it is also true for $n = k + 1$, and since it is true for $n = 1$, by induction we conclude that the result is true for any value of n, completing the proof. □

A.2 Full Proof of Proposition 3.4

We restate Proposition 3.4 here.

Proposition A.1. *Consider a graph Γ which contains exactly three cores induced by a complete family of ancestors $\mathbb{A}(\Gamma)$. Then suppose some sequence of inverse operations is performed such that each inverse operation preserves cores induced by $\mathbb{A}(\Gamma)$. Furthermore, after the sequence of inverse operations has been performed, there are no more inverse operations preserving cores induced by $\mathbb{A}(\Gamma)$ possible. Then,*

(i) the cores induced by $\mathbb{A}(\Gamma)$ have been separated, and
(ii) the resulting ancestral graph is precisely the ancestral graph associated with $\mathbb{A}(\Gamma)$.

Proof. By Corollary 3.1, if we show that the proposition holds for Γ containing no parthenogenic object then the proposition is true for all Γ. Therefore, we suppose that Γ contains no parthenogenic object. Denote by Γ_1, Γ_2, Γ_3 the cores of Γ induced by $\mathbb{A}(\Gamma)$. Let CE be the union of irreducible cracker edges that do not contain any edge in any of the three cores. The three cores in Γ could be connected to one another in a variety of ways. We will consider all possible configurations of the connections between the cores of Γ, which are found by exhaustive inspection in Proposition 3.3, and demonstrate that any method of disconnecting the cores through inverse operations results in $\mathbb{A}(\Gamma)$. Note that by Lemma 3.3 any sequence of inverse operations satisfying separating the three cores involves exactly two inverse breeding operations. We will go through all the cases in the following order. Cases 1–8 deal with situations in which a parthenogenic object cannot be created after the first inverse breeding operation, and so precisely two inverse operations are performed. Cases 9–15 deal with situations in which a parthenogenic object can be created after the first inverse breeding operation, and so more than two inverse operations may be required.

Case 1: There are two irreducible cubic crackers C_1 and C_2 which satisfy conditions (i)–(v) of Proposition 3.2. Then, by Proposition 3.2, the ancestors in $\mathbb{A}(\Gamma)$ can be found by performing inverse breeding operations on C_1 and C_2 in either order.

Case 2: There are two irreducible 1-crackers C_1, C_2 where C_1 and C_2 share an endpoint w. In this case the third edge connected to w, that is not in C_1 and C_2 must also be an irreducible 1-cracker. Without loss of generality suppose $CE = \{e_1 = (u_1, w), e_2 = (u_2, w), e_3 = (u_3, w)\}$ where $C_1 = \{e_1\}$, $C_2 = \{e_2\}$ and $C_3 = \{e_3\}$. Also u_1 is connected to $v_1, v_2 \in \Gamma_1$, u_2 is connected to $v_3, v_4 \in \Gamma_2$ and u_3 is connected to $v_5, v_6 \in \Gamma_3$. By considering all possible ways to proceed to perform inverse operations, we can verify that the outcome is always $\Gamma^{(2)} = \Gamma_1^{(2)} \cup \Gamma_2^{(2)} \cup \Gamma_3^{(2)}$, where

$$V(\Gamma_1^{(2)}) = V(\Gamma_1)$$
$$E(\Gamma_1^{(2)}) = E(\Gamma_1) \cup \{(v_1, v_2)\},$$

$$V(\Gamma_2^{(2)}) = V(\Gamma_2)$$
$$E(\Gamma_2^{(2)}) = E(\Gamma_2) \cup \{(v_3, v_4)\},$$

and

$$V(\Gamma_3^{(2)}) = V(\Gamma_3)$$
$$E(\Gamma_3^{(2)}) = E(\Gamma_3) \cup \{(v_5, v_6)\}.$$

This situation can be see in Fig. A.9.

Case 3: There are two irreducible 2-crackers C_1 and C_2 that share an edge e_1 and other edges in $C_1 \cup C_2$ do not share endpoints.

Without loss of generality suppose $CE = \{e_1 = (v_1, u_1), e_2 = (v_2, u_2), e_3 = (v_3, u_3)\}$, where $v_1, v_2 \in \Gamma_1$, $u_1, u_3 \in \Gamma_2$ and $u_2, v_3 \in \Gamma_3$. It is obvious that $C_1 = \{e_1, e_2\}$, $C_2 = \{e_1, e_3\}$ and $C_3 = \{e_2, e_3\}$ are all irreducible cubic crackers. By considering all possible ways to proceed to perform inverse breeding operations until all cores are disconnected, it can be checked that the outcome is always $\Gamma^{(2)} = \Gamma_1^{(2)} \cup \Gamma_2^{(2)} \cup \Gamma_3^{(2)}$, where

$$V(\Gamma_1^{(2)}) = V(\Gamma_1)$$
$$E(\Gamma_1^{(2)}) = E(\Gamma_1) \cup \{(v_1, v_2)\},$$

$$V(\Gamma_2^{(2)}) = V(\Gamma_2)$$
$$E(\Gamma_2^{(2)}) = E(\Gamma_2) \cup \{(u_1, u_3)\},$$

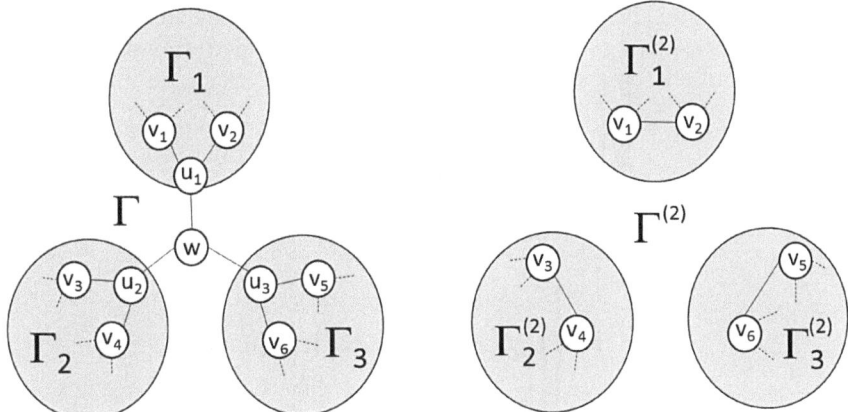

Fig. A.9 Graphs Γ and $\Gamma^{(2)}$ as described in Case 2 of Proposition 3.4

and

$$V(\Gamma_3^{(2)}) = V(\Gamma_3)$$
$$E(\Gamma_3^{(2)}) = E(\Gamma_3) \cup \{(v_3, u_2)\}.$$

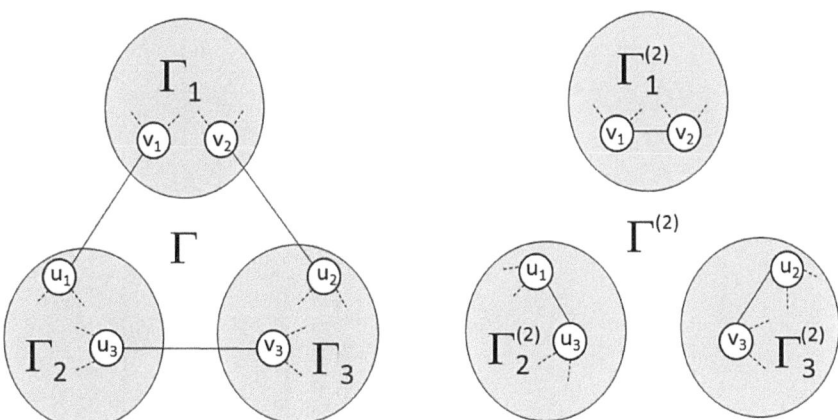

Fig. A.10 Graphs Γ and $\Gamma^{(2)}$ as described in Case 3 of Proposition 3.4

This situation can be see in Fig. A.10.

Case 4: There is a 2-cracker C_1 and a 3-cracker C_2, both irreducible, that share an edge e_1 and other edges in $C_1 \cup C_2$ do not share endpoints. In this case, the endpoint of cracker edges may or may not be distinct.

Without loss of generality suppose $CE = \{e_1 = (v_1, u_1), e_2 = (v_2, u_2), e_3 = (v_3, u_3), e_4 = (v_4, u_4)\}$ where $v_1, v_2 \in \Gamma_1$, $u_1, v_3, v_4 \in \Gamma_2$ and $u_2, u_3, u_4 \in \Gamma_3$. It is obvious that $C_1 = \{e_1, e_2\}$, $C_2 = \{e_1, e_3, e_4\}$ and $C_3 = \{e_2, e_3, e_4\}$ are all irreducible cubic crackers. By considering all possible ways to proceed to perform inverse breeding operations until all cores are disconnected, it can be checked that the outcome is always $\Gamma^{(2)} = \Gamma_1^{(2)} \cup \Gamma_2^{(2)} \cup \Gamma_3^{(2)}$, where

$$V(\Gamma_1^{(2)}) = V(\Gamma_1)$$
$$E(\Gamma_1^{(2)}) = E(\Gamma_1) \cup \{(v_1, v_2)\},$$

$$V(\Gamma_2^{(2)}) = V(\Gamma_2) \cup \{w_1\}$$
$$E(\Gamma_2^{(2)}) = E(\Gamma_2) \cup \{(u_1, w_1), (v_3, w_1), (v_4, w_1)\},$$

and

$$V(\Gamma_3^{(2)}) = V(\Gamma_3) \cup \{w_2\}$$
$$E(\Gamma_3^{(2)}) = E(\Gamma_3) \cup \{(u_2, w_2), (u_3, w_2), (u_4, w_2)\},$$

where w_1 and w_2 are two new vertices. This situation can be seen in Fig. A.11. Note that this case also covers the situation where two three crackers (C_2 and C_3 in this example) share two edges, so this case will not be considered further.

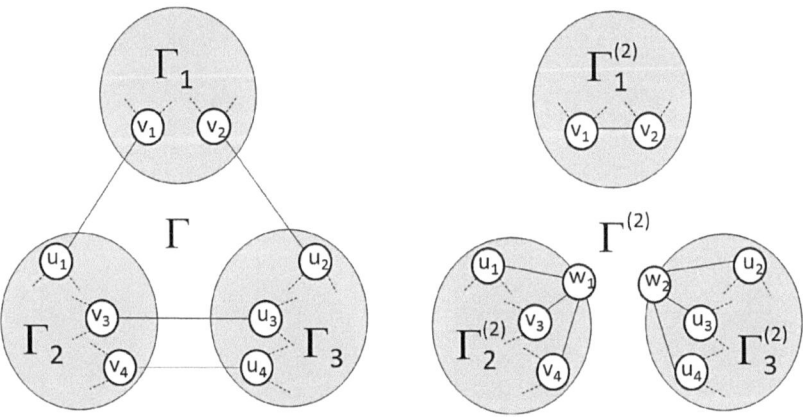

Fig. A.11 Graphs Γ and $\Gamma^{(2)}$ as described in Case 4 of Proposition 3.4

Case 5: There are two irreducible 3-crackers C_1 and C_2 that share an edge e_1.

Without loss of generality suppose $CE = \{e_1 = (v_1, u_1), e_2 = (v_2, u_2), e_3 = (v_3, u_3), e_4 = (v_4, u_4), e_5 = (v_5, u_5)\}$ where $u_2, u_3, u_4, u_5 \in \Gamma_1$, $v_1, v_2, v_3 \in \Gamma_2$ and $u_1, v_4, v_5 \in \Gamma_3$. It is obvious that $C_1 = \{e_1, e_2, e_3\}$, $C_2 = \{e_1, e_4, e_5\}$ are all irreducible cubic crackers. By considering all possible ways to proceed to perform inverse breeding operations until all cores are disconnected, it can be checked that the outcome is always $\Gamma^{(2)} = \Gamma_1^{(2)} \cup \Gamma_2^{(2)} \cup \Gamma_3^{(2)}$, where

$$V(\Gamma_1^{(2)}) = V(\Gamma_1) \cup \{w_1, w_2\}$$
$$E(\Gamma_1^{(2)}) = E(\Gamma_1) \cup \{(u_2, w_1), (u_3, w_1), (u_4, w_2), (u_5, w_2), (w_1, w_2)\},$$

$$V(\Gamma_2^{(2)}) = V(\Gamma_2) \cup \{w_3\}$$
$$E(\Gamma_2^{(2)}) = E(\Gamma_2) \cup \{(v_1, w_3), (v_2, w_3), (v_3, w_3)\},$$

and

$$V(\Gamma_3^{(2)}) = V(\Gamma_3) \cup \{w_4\}$$
$$E(\Gamma_3^{(2)}) = E(\Gamma_3) \cup \{(u_1, w_4), (v_4, w_4), (v_5, w_4)\},$$

where w_1, w_2, w_3 and w_4 are new vertices. This situation is displayed in Fig. A.12.

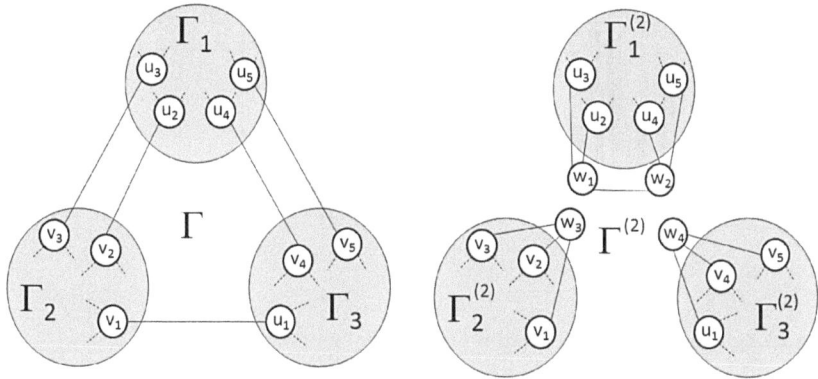

Fig. A.12 Graphs Γ and $\Gamma^{(2)}$ as described in Case 5 of Proposition 3.4

Case 6: There is an edge $e = (b_1, b_2)$ adjacent to two irreducible 1-crackers, C_1 and C_2.

Without loss of generality suppose $CE = \{e_1 = (b_1, b_3), e_2 = (b_2, b_4)\}$, where $C_1 = \{e_1\}$ and $C_2 = \{e_2\}$ are both irreducible cubic crackers. Also the vertices a_1, b_2 are adjacent to b_1, the vertices a_2, b_1 are adjacent to b_2, the vertices a_3, a_4 are adjacent to b_3 and the vertices a_5, a_6 are adjacent to b_4. Furthermore, $a_1, a_2 \in \Gamma_1$, $a_3, a_4 \in \Gamma_2$ and $a_5, a_6 \in \Gamma_3$. By considering all possible ways to proceed to perform inverse breeding operations until all cores are disconnected, it can be checked that the outcome is always $\Gamma^{(2)} = \Gamma_1^{(2)} \cup \Gamma_2^{(2)} \cup \Gamma_3^{(2)}$, where

$$V(\Gamma_1^{(2)}) = V(\Gamma_1)$$
$$E(\Gamma_1^{(2)}) = E(\Gamma_1) \cup \{(a_1, a_2)\},$$

$$V(\Gamma_2^{(2)}) = V(\Gamma_2)$$
$$E(\Gamma_2^{(2)}) = E(\Gamma_2) \cup \{(a_3, a_4)\},$$

and

$$V(\Gamma_3^{(2)}) = V(\Gamma_3)$$
$$E(\Gamma_3^{(2)}) = E(\Gamma_3) \cup \{(a_5, a_6)\}.$$

This situation is displayed in Fig. A.13.

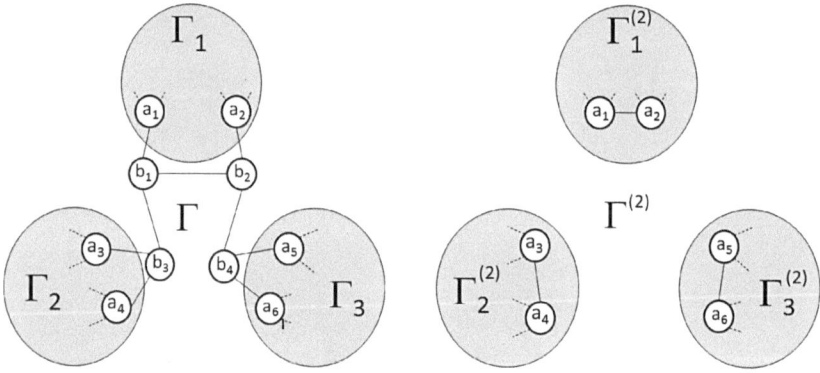

Fig. A.13 Graphs Γ and $\Gamma^{(2)}$ as described in Case 6 of Proposition 3.4

Case 7: There exists an edge e_1 adjacent to an irreducible 1-cracker C_1, such that e_1 is an edge of an irreducible 2-cracker C_2.

Without loss of generality suppose $CE = \{e_1 = (a_4, b_2), e_2 = (b_1, b_2), e_3 = (v_1, u_1), e_4 = (b_2, a_3)\}$, where $C_1 = \{e_2\}$, $C_2 = \{e_1, e_3\}$ and $C_3 = \{e_3, e_4\}$ are all irreducible cubic crackers. Also the vertices a_1, a_2 are adjacent to b_1. Furthermore, $a_1, a_2 \in \Gamma_1$, $a_3, v_1 \in \Gamma_2$ and $a_4, u_1 \in \Gamma_3$. By considering all possible ways to proceed to perform inverse breeding operations until all cores are disconnected, it can be checked that the outcome is always $\Gamma^{(2)} = \Gamma_1^{(2)} \cup \Gamma_2^{(2)} \cup \Gamma_3^{(2)}$, where

$$V(\Gamma_1^{(2)}) = V(\Gamma_1)$$
$$E(\Gamma_1^{(2)}) = E(\Gamma_1) \cup \{(a_1, a_2)\},$$

$$V(\Gamma_2^{(2)}) = V(\Gamma_2)$$
$$E(\Gamma_2^{(2)}) = E(\Gamma_2) \cup \{(a_3, v_1)\},$$

and

$$V(\Gamma_3^{(2)}) = V(\Gamma_3)$$
$$E(\Gamma_3^{(2)}) = E(\Gamma_3) \cup \{(a_4, u_1)\}.$$

This situation is displayed in Fig. A.14.

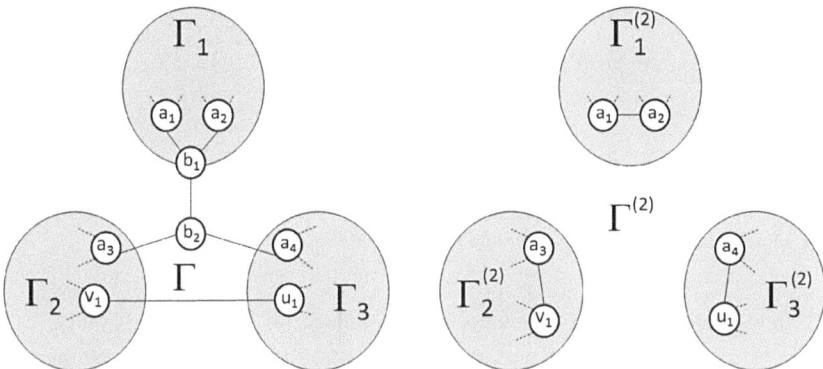

Fig. A.14 Graphs Γ and $\Gamma^{(2)}$ as described in Case 7 of Proposition 3.4

Case 8: There is an edge e_1 adjacent to an irreducible bridge C_1, such that e_1 is an edge of an irreducible 3-cracker C_2.

Without loss of generality suppose $CE = \{e_1 = (a_4, b_2), e_2 = (b_1, b_2), e_3 = (v_1, u_1), e_4 = (v_2, u_2), e_5 = (b_2, a_3)\}$ where $C_1 = \{e_2\}$, $C_2 = \{e_1, e_3, e_4\}$ and $C_3 = \{e_3, e_4, e_5\}$ are all irreducible cubic crackers. Also the vertices a_1, a_2 are adjacent to b_1. Furthermore, $a_1, a_2 \in \Gamma_1$, $a_3, v_1, v_2 \in \Gamma_2$ and $a_4, u_1, u_2 \in \Gamma_3$. By considering all possible ways to proceed to perform inverse breeding operations until all cores are disconnected, it can be checked that the outcome is always $\Gamma^{(2)} = \Gamma_1^{(2)} \cup \Gamma_2^{(2)} \cup \Gamma_3^{(2)}$, where

$$V(\Gamma_1^{(2)}) = V(\Gamma_1)$$
$$E(\Gamma_1^{(2)}) = E(\Gamma_1) \cup \{(a_1, a_2)\},$$

$$V(\Gamma_2^{(2)}) = V(\Gamma_2) \cup \{w_1\}$$
$$E(\Gamma_2^{(2)}) = E(\Gamma_2) \cup \{(a_3, w_1), (v_1, w_1), (v_2, w_1)\},$$

and

$$V(\Gamma_3^{(2)}) = V(\Gamma_3) \cup \{w_2\}$$
$$E(\Gamma_3^{(2)}) = E(\Gamma_3) \cup \{(a_4, w_2), (u_1, w_2), (u_2, w_2)\},$$

where w_1 and w_2 are two new vertices. This situation can be seen in Fig. A.15.

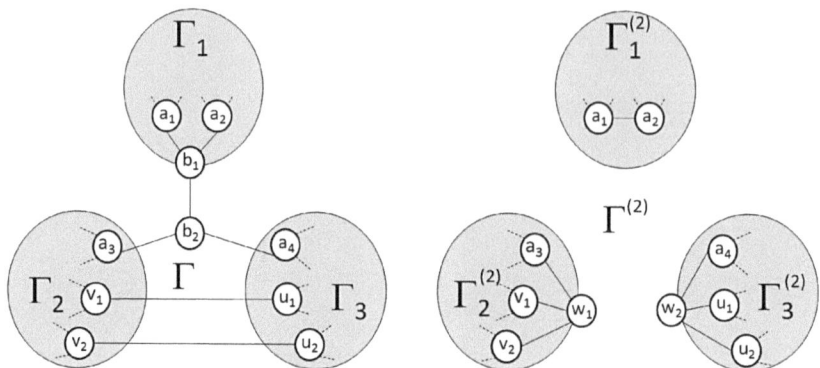

Fig. A.15 Graphs Γ and $\Gamma^{(2)}$ as described in Case 8 of Proposition 3.4

Case 9: There exists an inverse type 1 breeding operation applicable on Γ such that the resultant graph has a parthenogenic diamond which is not contained in the cores induced by $\mathbb{A}(\Gamma)$.

Without loss of generality, suppose the inverse type 3 operation is performed on $C_1 = \{(b_7, d_3)\}$ where the vertices $b_3, b_4, b_5, b_6 \in V(\Gamma)$ will be three of the four vertices of the parthenogenic diamond. The union of irreducible crackers $CE = C_1 \cup C_2 \cup C_3 = \{(b_7, d_3), (b_1, b_3), (b_2, b_5)\}$, where $C_2 = \{(b_1, b_3)\}$ and $C_3 = \{(b_5, b_2)\}$ are other irreducible cubic crackers in CE. Also the vertices a_1, a_2 are adjacent to b_1, the vertices a_3, a_4 are adjacent to b_2 and vertices d_1, d_2 are adjacent to d_3. Furthermore, $d_1, d_2 \in \Gamma_1$, $a_1, a_2 \in \Gamma_2$ and $a_3, a_4 \in \Gamma_3$. By considering all possible ways to proceed to perform inverse breeding operations until all cores are disconnected, it can be checked that the outcome is always $\Gamma^{(2)} = \Gamma_1^{(2)} \cup \Gamma_2^{(2)} \cup \Gamma_3^{(2)}$, where

$$V(\Gamma_1^{(2)}) = V(\Gamma_1)$$
$$E(\Gamma_1^{(2)}) = E(\Gamma_1) \cup \{(d_1, d_2)\},$$

$$V(\Gamma_2^{(2)}) = V(\Gamma_2)$$
$$E(\Gamma_2^{(2)}) = E(\Gamma_2) \cup \{(a_1, a_2)\},$$

and

$$V(\Gamma_3^{(2)}) = V(\Gamma_3)$$
$$E(\Gamma_3^{(2)}) = E(\Gamma_3) \cup \{(a_3, a_4)\}.$$

This situation can be seen in Fig. A.16.

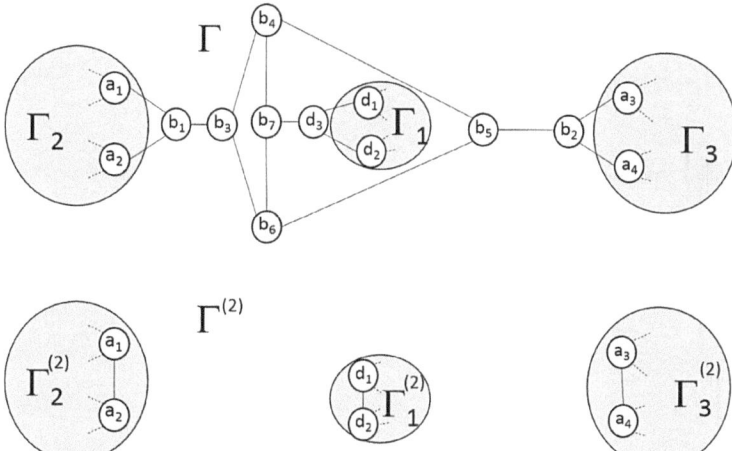

Fig. A.16 Graphs Γ and $\Gamma^{(2)}$ as described in Case 9 of Proposition 3.4

Case 10: There exists an inverse type 2 breeding operation applicable on Γ such that the resultant graph contains a parthenogenic diamond which is not contained in the cores induced by $\mathbb{A}(\Gamma)$.

Without loss of generality, suppose that when an inverse type 2 breeding operation is performed on $C_1 = \{(d_1, b_4), (d_2, b_6)\}$, a parthenogenic diamond is formed containing vertices $b_3, b_4, b_5, b_6 \in V(\Gamma)$. The union of crackers $CE = C_1 \cup C_2 \cup C_3 = \{(d_1, b_4), (d_2, b_6), (b_1, b_3), (b_2, b_5)\}$, where $C_2 = \{(b_1, b_3)\}$ and $C_3 = \{(b_2, b_5)\}$ are the other irreducible cubic crackers in CE. Also the vertices a_1, a_2 are adjacent to b_1, and the vertices a_3, a_4 are adjacent to b_2. Furthermore, $d_1, d_2 \in \Gamma_1$, $a_1, a_2 \in \Gamma_2$ and $a_3, a_4 \in \Gamma_3$. By considering all possible ways to proceed to perform inverse breeding operations until all cores are disconnected, it can be checked that the outcome is always $\Gamma^{(2)} = \Gamma_1^{(2)} \cup \Gamma_2^{(2)} \cup \Gamma_3^{(2)}$, where

$$V(\Gamma_1^{(2)}) = V(\Gamma_1)$$
$$E(\Gamma_1^{(2)}) = E(\Gamma_1) \cup \{(d_1, d_2)\},$$

$$V(\Gamma_2^{(2)}) = V(\Gamma_2)$$
$$E(\Gamma_2^{(2)}) = E(\Gamma_2) \cup \{(a_1, a_2)\},$$

and

$$V(\Gamma_3^{(2)}) = V(\Gamma_3)$$
$$E(\Gamma_3^{(2)}) = E(\Gamma_3) \cup \{(a_3, a_4)\}.$$

This situation is displayed in Fig. A.17.

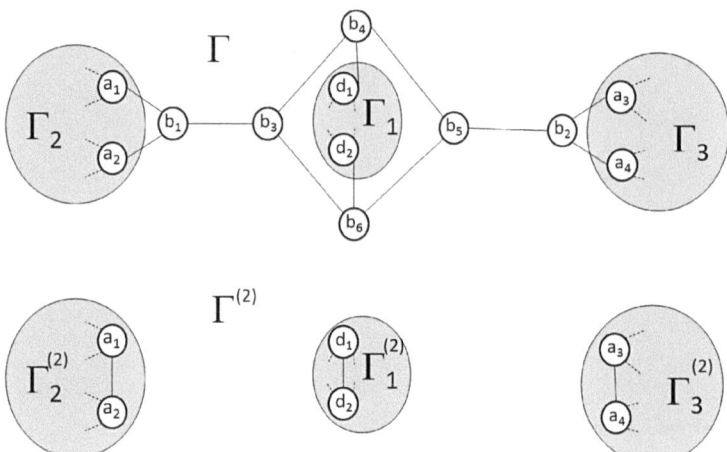

Fig. A.17 Graphs Γ and $\Gamma^{(2)}$ as described in Case 10 of Proposition 3.4

Case 11: There exists an inverse type 3 breeding operation applicable on Γ such that the resultant graph has a parthenogenic diamond which is not contained in the cores induced by $\mathbb{A}(\Gamma)$.

Without loss of generality, suppose the inverse type 3 operation is performed on $C_1 = \{(d_1,b_3),(d_2,b_4),(d_3,b_5)\}$ where the vertices $b_3,b_4,b_5 \in V(\Gamma)$ will be three of the four vertices of the parthenogenic diamond that results. The fourth vertex of the parthenogenic diamond is created by the inverse operation C_1. The union of crackers $CE = C_1 \cup C_2 \cup C_3 = \{(d_1,b_3),(d_2,b_4),(d_3,b_5),(b_1,b_3),(b_2,b_5)\}$, where $C_2 = \{(b_1,b_3)\}$ and $C_3 = \{(b_2,b_5)\}$ are the other irreducible cubic crackers in CE. Also the vertices a_1,a_2 are adjacent to b_1, and the vertices a_3,a_4 are adjacent to b_2. Furthermore, $d_1,d_2,d_3 \in \Gamma_1$, $a_1,a_2 \in \Gamma_2$ and $a_3,a_4 \in \Gamma_3$. By considering all possible ways to proceed to perform inverse breeding operations until all cores are disconnected, it can be checked that the outcome is always $\Gamma^{(2)} = \Gamma_1^{(2)} \cup \Gamma_2^{(2)} \cup \Gamma_3^{(2)}$, where

$$V(\Gamma_1^{(2)}) = V(\Gamma_1) \cup \{w\}$$
$$E(\Gamma_1^{(2)}) = E(\Gamma_1) \cup \{(d_1,w),(d_2,w),(d_3,w)\},$$
$$V(\Gamma_2^{(2)}) = V(\Gamma_2)$$
$$E(\Gamma_2^{(2)}) = E(\Gamma_2) \cup \{(a_1,a_2)\},$$

and

$$V(\Gamma_3^{(2)}) = V(\Gamma_3)$$
$$E(\Gamma_3^{(2)}) = E(\Gamma_3) \cup \{(a_3,a_4)\},$$

where w is a new vertex. This situation is displayed in Fig. A.18.

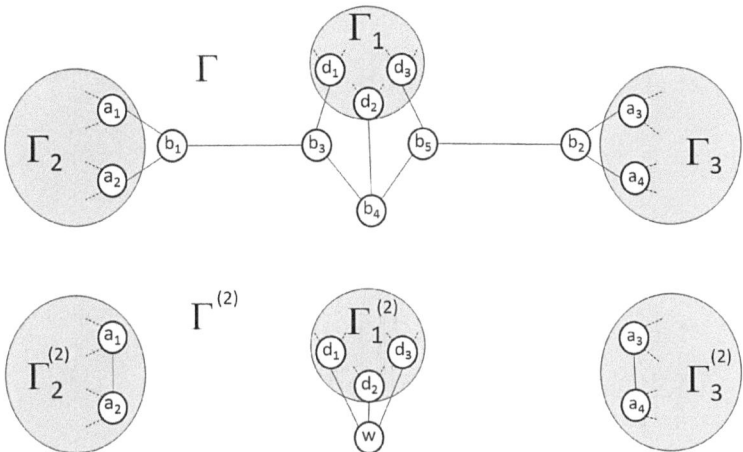

Fig. A.18 Graphs Γ and $\Gamma^{(2)}$ as described in Case 11 of Proposition 3.4

Case 12: There exists an inverse type 2 breeding operation applicable on Γ such that the resultant graph has a parthenogenic bridge which is not contained in the cores induced by $\mathbb{A}(\Gamma)$.

Without loss of generality, suppose that when an inverse type 2 breeding operation is performed on $C_1 = \{(d_1, b_1), (d_2, b_2)\}$, a parthenogenic bridge is formed consisting of the vertices $b_1, b_2 \in V(\Gamma)$. The union of crackers $CE = C_1 \cup C_2 \cup C_3 = \{(d_1, b_1), (d_2, b_2), (a_1, b_1), (a_2, b_2), (a_3, b_1), (a_4, b_2)\}$, where $C_2 = \{(a_1, b_1), (a_2, b_2)\}$ and $C_2 = \{(a_3, b_1), (a_4, b_2)\}$ are the other irreducible cubic crackers in CE. Furthermore, $d_1, d_2 \in \Gamma_1$, $a_1, a_2 \in \Gamma_2$ and $a_3, a_4 \in \Gamma_3$. By considering all possible ways to proceed to perform inverse breeding operations until all cores are disconnected, it can be checked that the outcome is always $\Gamma^{(2)} = \Gamma_1^{(2)} \cup \Gamma_2^{(2)} \cup \Gamma_3^{(2)}$, where

$$V(\Gamma_1^{(2)}) = V(\Gamma_1)$$
$$E(\Gamma_1^{(2)}) = E(\Gamma_1) \cup \{(d_1, d_2)\},$$

$$V(\Gamma_2^{(2)}) = V(\Gamma_2)$$
$$E(\Gamma_2^{(2)}) = E(\Gamma_2) \cup \{(a_1, a_2)\},$$

and

$$V(\Gamma_3^{(2)}) = V(\Gamma_3)$$
$$E(\Gamma_3^{(2)}) = E(\Gamma_3) \cup \{(a_3, a_4)\}.$$

This situation id displayed in Fig. A.19.

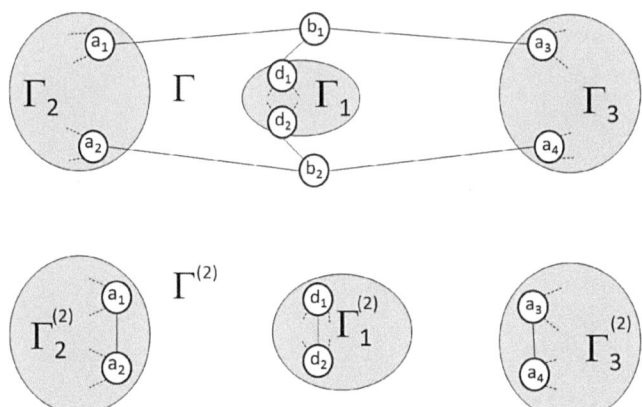

Fig. A.19 Graphs Γ and $\Gamma^{(2)}$ as described in Case 12 of Proposition 3.4

Case 13: There exists an inverse type 3 breeding operation applicable on Γ such that the resultant graph has a parthenogenic bridge which is not contained in the cores induced by $\mathbb{A}(\Gamma)$.

Without loss of generality, suppose that when an inverse type 3 breeding operation is performed on $C_1 = \{(d_1,a_1),(d_2,b_1),(d_3,a_3)\}$, a parthenogenic bridge is formed consisting of the vertex $b_1 \in V(\Gamma)$, and a vertex created by the inverse type 3 breeding operation. The union of crackers $CE = C_1 \cup C_2 \cup C_3 = \{(d_1,a_1),(d_2,b_1),$ $(d_3,a_3),(a_2,b_1),(b_1,a_4)\}$, where $C_2 = \{(a_1,d_1),(a_2,b_1)\}$ and $C_3 = \{(a_3,d_3),$ $(a_4,b_1)\}$ are the other irreducible cubic crackers in CE. Furthermore, $d_1,d_2,d_3 \in \Gamma_1$, $a_1,a_2 \in \Gamma_2$ and $a_3,a_4 \in \Gamma_3$. By considering all possible ways to proceed to perform inverse breeding operations until all cores are disconnected, it can be checked that the outcome is always $\Gamma^{(2)} = \Gamma_1^{(2)} \cup \Gamma_2^{(2)} \cup \Gamma_3^{(2)}$, where

$$V(\Gamma_1^{(2)}) = V(\Gamma_1) \cup \{w\}$$
$$E(\Gamma_1^{(2)}) = E(\Gamma_1) \cup \{(d_1,w),(d_2,w),(d_3,w)\},$$

$$V(\Gamma_2^{(2)}) = V(\Gamma_2)$$
$$E(\Gamma_2^{(2)}) = E(\Gamma_2) \cup \{(a_1,a_2)\},$$

and

$$V(\Gamma_3^{(2)}) = V(\Gamma_3)$$
$$E(\Gamma_3^{(2)}) = E(\Gamma_3) \cup \{(a_3,a_4)\},$$

where w is a new vertex. This situation is displayed in Fig. A.20.

Case 14: There exists an inverse type 2 breeding operation applicable on Γ such that the resultant graph has a parthenogenic triangle which is not contained in the cores induced by $\mathbb{A}(\Gamma)$.

Without loss of generality, suppose that when an inverse type 2 breeding is performed on $C_1 = \{(d_1, b_2), (d_2, b_3)\}$, a parthenogenic triangle is formed consisting of vertices $b_2, b_3, b_4 \in V(\Gamma)$. The union of crackers $CE = C_1 \cup C_2 \cup C_3 = \{(d_1, b_2), (d_2, b_3), (a_1, b_2), (a_2, b_3), (b_1, b_4)\}$, where $C_2 = \{(b_1, b_4)\}$ and $C_3 = \{(a_1, b_2), (a_2, b_3)\}$ are the other irreducible cubic crackers in CE. Also the vertices a_3, a_4 are adjacent to b_1. Furthermore, $d_1, d_2 \in \Gamma_1$, $a_1, a_2 \in \Gamma_2$ and $a_3, a_4 \in \Gamma_3$. By considering all possible ways to proceed to perform inverse breeding operations until all cores are disconnected, it can be checked that the outcome is always $\Gamma^{(2)} = \Gamma_1^{(2)} \cup \Gamma_2^{(2)} \cup \Gamma_3^{(2)}$, where

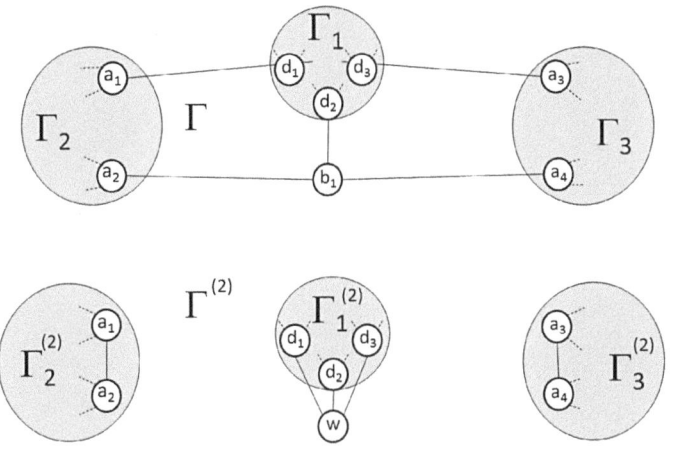

Fig. A.20 Graphs Γ and $\Gamma^{(2)}$ as described in Case 13 of Proposition 3.4

$$V(\Gamma_1^{(2)}) = V(\Gamma_1)$$
$$E(\Gamma_1^{(2)}) = E(\Gamma_1) \cup \{(d_1, d_2)\},$$

$$V(\Gamma_2^{(2)}) = V(\Gamma_2)$$
$$E(\Gamma_2^{(2)}) = E(\Gamma_2) \cup \{(a_1, a_2)\},$$

and

$$V(\Gamma_3^{(2)}) = V(\Gamma_3)$$
$$E(\Gamma_3^{(2)}) = E(\Gamma_3) \cup \{(a_3, a_4)\}.$$

This situation is displayed in Fig. A.21.

Case 15: There exists an inverse type 3 breeding operation applicable on Γ such that the resultant graph has a parthenogenic triangle which is not contained in the cores induced by $\mathbb{A}(\Gamma)$.

Without loss of generality, suppose that when an inverse type 3 operation is performed on $C_1 = \{(d_1,a_1),(d_2,b_3),(d_3,b_2)\}$, a parthenogenic triangle is formed consisting of vertices $b_2,b_3 \in V(\Gamma)$, and a third vertex created by the inverse type 3 breeding operation. The union of crackers $CE = C_1 \cup C_2 \cup C_3 = \{(d_1,a_1),(d_2,b_3),$ $(d_3,b_2),(a_2,b_3),(b_1,b_2)\}$, where $C_2 = \{(b_1,b_2)\}$ and $C_3 = \{(a_1,d_1),(a_2,b_3)\}$ are the other irreducible cubic crackers in CE. Also the vertices a_3,a_4 are adjacent to b_1. Furthermore, $d_1,d_2,d_3 \in \Gamma_1$, $a_1,a_2 \in \Gamma_2$ and $a_3,a_4 \in \Gamma_3$. By considering all possible ways to proceed to perform inverse breeding operations until all cores are disconnected, it can be checked that the outcome is always $\Gamma^{(2)} = \Gamma_1^{(2)} \cup \Gamma_2^{(2)} \cup \Gamma_3^{(2)}$, where

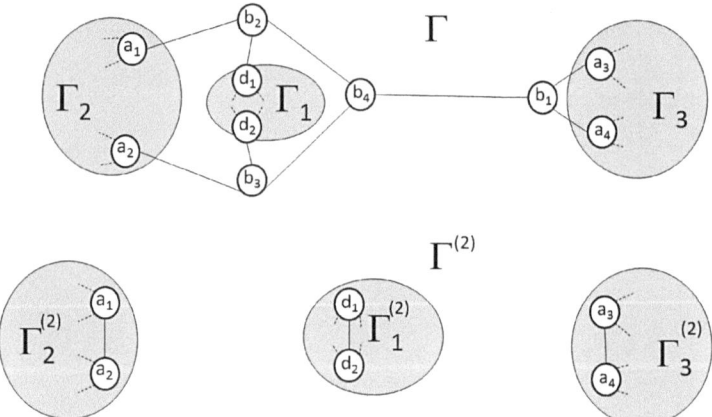

Fig. A.21 Graphs Γ and $\Gamma^{(2)}$ as described in Case 14 of Proposition 3.4

$$V(\Gamma_1^{(2)}) = V(\Gamma_1) \cup \{w\}$$
$$E(\Gamma_1^{(2)}) = E(\Gamma_1) \cup \{(d_1,w),(d_2,w),(d_3,w)\},$$

$$V(\Gamma_2^{(2)}) = V(\Gamma_2)$$
$$E(\Gamma_2^{(2)}) = E(\Gamma_2) \cup \{(a_1,a_2)\},$$

and

$$V(\Gamma_3^{(2)}) = V(\Gamma_3)$$
$$E(\Gamma_3^{(2)}) = E(\Gamma_3) \cup \{(a_3,a_4)\},$$

where w is a new vertex. This situation is displayed in Fig. A.22.

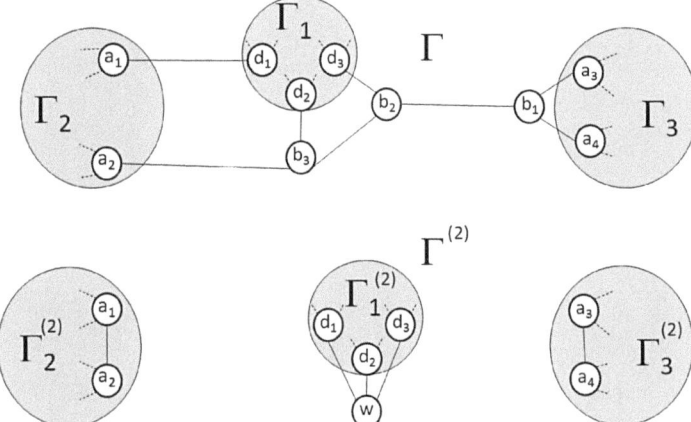

Fig. A.22 Graphs Γ and $\Gamma^{(2)}$ as described in Case 15 of Proposition 3.4

Since we have considered all 15 cases of Proposition 3.3, the proof is complete. □

Index